PRAISE F

Lost Moun

"With a journalist's eye, a naturalist's h
ing engineer's son, Reece produces a powerful environmental exposé, documenting the process on one sorry Kentucky peak."

—*Outside* magazine

"Orwell and Kafka in their bleakest moments would have felt right at home in Appalachian Kentucky, mired in corruption and class warfare. . . . Written with an eye for abiding, catastrophic imagery . . . a portrait of coal country as stark and galvanizing as Harry Caudill's classic *Night Comes to the Cumberlands.*" —*The Tampa Tribune*

"*Lost Mountain* is a great book arriving at a crucial time. Mountaintop removal is a crime against humanity. The terrible irony is that the people who should most read the book won't open it—the ones destroying the land and taking the profits out of state. It's up to you to read *Lost Mountain* and voice your concerns, not just for Kentucky, but for the future of America. Appalachia is everyone's backyard. Your front yard is next."

—Chris Offutt, author of *Kentucky Straight*

"Finally a book has been written about mountaintop removal. While Kentucky's and West Virginia's Appalachian peaks are being flattened to the tune of 1,955 square miles (equivalent to the state of Delaware), America has stood by in shameful ignorance and disregard. Now Erik Reece has lifted the veil from the mining industry's preferred method of extracting coal: quickly, cheaply, with minimum manpower, and maximum ecological destruction. . . . This beautifully written book is also a jeremiad to an entire way of life in Appalachia—a life closely linked to nature and the mountains—which is on the verge of being wiped out by America's addiction to coal."

—*The Louisville Courier-Journal*

"In compelling prose peppered with cold, hard facts, [Reece] tracked the fate of one Appalachian peak, aptly named Lost Mountain. . . . Lost Mountain was so named because early hunters often became disoriented among its dense and verdant trees. As Reece conveys with superb reporting, we still have not found our way."

—*Discover* magazine

continued . . .

"*Lost Mountain* is a story well told, both eloquent and moving. It is a requiem for 'a landscape worthy of comparison to an earthly paradise.' . . . But this is not just a tale of environmental ruination. The blasting, coal washing, and valley filling create deep human suffering, raising issues of decency, fairness, and justice. . . . But no good book is purely a lament, and this one certainly is not. Instead it looks forward to a better era when at last we will have built a culture to match the peaks of Appalachia. And we have neither time nor mountains to waste."
—*American Scientist*

"*Lost Mountain* is a documentary, a personal narrative, a natural history of the Lost Mountain ecosystem. It's also a social history of Appalachia, an evaluation of its modern political significance, and always, always a green threnody."
—*The Buffalo News*

"Time to listen to the words of American ground. This eloquent book tells us what we already know in our bones and will soon have slapped in our face: We can't murder mountains without killing ourselves. Decades after Appalachia became a symbol of American failure and shame, we seem to have forgotten the lessons our fathers and mothers taught us. In *Lost Mountain*, Erik Reece shows us the way to go home."
—Charles Bowden, author of *Blood Orchid*

"Reece's report is a powerful indictment of the lax oversight of mining regulations and their scuttling by political allies of the mining industry . . . an indictment made even more powerful by Reece's warmhearted accounts of time spent with the Perry County families whose health and homes have been threatened by the mine's operation, and by his graceful descriptions of the natural splendor of eastern Kentucky. . . . There are echoes of Rachel Carson's warning of ecological disaster in *Lost Mountain*, but perhaps the deeper tragedy is that so little has changed in eastern Kentucky since Harry Caudill's searing account of the region's social and economic poverty, *Night Comes to the Cumberlands*, from 1963."
—*The Boston Globe*

"Reece writes at times with a naturalist's lyric ease, at others with the urgency of the activist's call to arms. . . . Indeed, by the end of this careful and illuminating chronicle of destruction, *Lost Mountain* begins to represent all that is heedlessly selfish in America's attitude toward its environment."
—*The Virginia Quarterly Review*

"For all its pleasures, we don't come to so-called nature writing expecting a 'gripping read.' But I couldn't put down *Lost Mountain*. There's a dissonance in the book—between some extremely ugly environmental realities that Erik Reece has worked hard to understand and the superb, unpretentious prose in which he describes them—that kept me pinned. And of course, apart from its stylistic virtues, *Lost Mountain* happens to be important."

—John Jeremiah Sullivan, author of *Blood Horses*

"Compelling, considered, and courageous . . . Read this book and take action." —*Library Journal*

"A searing indictment of how a country's energy lust is ravaging the hills and hollows of Appalachia. [An] elegiac book—much more than just an eyewitness report on ecological decimation . . . The Kentucky-born author, who canoed clean Appalachian rivers as a youth, has written an impassioned account of a business rife with industrial greed, devious corporate ownership, and unenforced environmental laws. It's also a heartrending account of the rural residents whose lives are being ruined by strip mining's relentless, almost unfettered, encroachment." —*Publishers Weekly*

"A portrait of coal country as stark and galvanizing as Harry Caudill's classic *Night Comes to the Cumberlands.*" —*Kirkus Reviews*

"Criminal. That's the word that comes to mind upon reading Reece's excoriating exposé of the coal industry's pernicious rape of the mountains of eastern Kentucky. Once the site of the oldest and most ecologically diverse forest in the country, now this stretch of Appalachian wilderness has gone from being a verdant North American rain forest to a bleak and dismal lunar landscape. . . . The tale of Kentucky's mutilated environment is one that, like the mountain, has been lost. Resounding kudos to Reece for vividly bringing this critical story to light." —*Booklist* (starred review)

LOST MOUNTAIN

A Year in the Vanishing Wilderness

Radical Strip Mining and the Devastation of Appalachia

ERIK REECE

Foreword by Wendell Berry

Photographs by John J. Cox

RIVERHEAD BOOKS

New York

THE BERKLEY PUBLISHING GROUP
Published by the Penguin Group
Penguin Group (USA) Inc.
375 Hudson Street, New York, New York 10014, USA
Penguin Group (Canada), 90 Eglinton Avenue East, Suite 700, Toronto, Ontario M4P 2Y3, Canada
(a division of Pearson Penguin Canada Inc.)
Penguin Books Ltd., 80 Strand, London WC2R 0RL, England
Penguin Group Ireland, 25 St. Stephen's Green, Dublin 2, Ireland (a division of Penguin Books Ltd.)
Penguin Group (Australia), 250 Camberwell Road, Camberwell, Victoria 3124, Australia
(a division of Pearson Australia Group Pty. Ltd.)
Penguin Books India Pvt. Ltd., 11 Community Centre, Panchsheel Park, New Delhi—110 017, India
Penguin Group (NZ), Cnr Airborne and Rosedale Roads, Albany, Auckland 1310, New Zealand
(a division of Pearson New Zealand Ltd.)
Penguin Books (South Africa) (Pty.) Ltd., 24 Sturdee Avenue, Rosebank, Johannesburg 2196,
South Africa

Penguin Books Ltd., Registered Offices: 80 Strand, London WC2R 0RL, England

Book design by Meighan Cavanaugh

Portions of this book originally appeared in *Harper's Magazine*, often in significantly different form.

The author acknowledges permission to quote from the following:

"I Am a Union Woman" by Aunt Molly Jackson, copyright © 1960
(renewed) by Stormking Music, Inc. All rights reserved. Used by permission.

"Manifesto: The Mad Farmer Liberation Front" by Wendell Berry. Reprinted by permission of the author.

"Which Side Are You On?" by Florence Reece, copyright © 1947
(renewed) by Stormking Music, Inc. All rights reserved. Used by permission.

"Sixteen Tons" by Merle Travis. Reprinted by permission of Merle's Girls Music.

"Home for the Weekend" by Gurney Norman. Reprinted from
Kinfolks: The Wilgus Stories by permission of Gnomon Press.

First Riverhead hardcover edition: February 2006
First Riverhead trade paperback edition: February 2007
Riverhead trade paperback ISBN: 978-1-59448-236-6

The Library of Congress has catalogued the Riverhead hardcover edition as follows:

Reece, Erik.
Lost mountain : a year in the vanishing wilderness :
radical strip mining and the devastation of Appalachia / Erik Reece.
p. cm.
Includes bibliographical references.
ISBN 1-59448-908-4
1. Strip mining—Environmental aspects—Appalachian Region. I. Title.
TD195.S75R43 2006 2005052921
622'.2920974—dc22

PRINTED IN THE UNITED STATES OF AMERICA

10 9 8 7 6 5 4 3 2 1

For Scott, Jenny, and Carson

Contents

PART TWO

Every valley shall be filled and
every mountain and hill brought low.

LUKE 3:5

Foreword

Most people, even the many whose homes are cooled, heated, and lighted by coal-fired power plants, know little or nothing about surface mining in the Appalachian coal fields. They know so little, for one reason, because this is a subject hard to learn about. The coal companies, knowing well what an abomination surface mining is, have gone to considerable trouble to hide it from public view. The "media" have paid it far too little attention, even though it is a matter of the most urgent public interest.

Another reason for so much ignorance is that the learning is painful. To know about strip mining or mountaintop removal is like knowing about the nuclear bomb. It is to know beyond doubt that some human beings have, and are willing to use, the power of absolute destruction. This work is done in violation of all the best things that humans have learned in their long dwelling on the earth: reverence, neighborliness, stewardship, thrift, love.

A conservationist trying to oppose this enormity must accept heartbreak as a working condition. People whose homes and homelands are under the dominance of the coal industry must accept heartbreak, poverty, and various everyday lethal endangerments as a way of life. And the coal companies themselves must live always a little frantically, trying to protect their contempt for everything human and natural by "Private Property" signs and purchased political friendships.

So much needs to be said as a way of suggesting our inestimable debt to Erik Reece for writing this book. This is by far the best account of mountaintop removal and of its effects. It is a superb job of reporting, and we have it at the cost of the effort, grief, and risk involved in observing from beginning to end the process of the industrial destruction of a mountain and the ruin of its watersheds. No other reporter has had the perseverance and the guts to do a respectable fraction of what Mr. Reece has done.

He has worked in the face of a public ignorance both conventional and enforced. His book confronts a disgraceful history in which generations of political hirelings have sacrificed their land and their people to the benefit of a few mainly absentee corporations. And so the Appalachian coal fields most spectacularly, but in fact every one of our economic landscapes, have been put at the mercy of a class of economic aggressors whose aim is to convert the natural world into money as quickly as is technologically possible and at the least possible cost. If that least cost is the total destruction of the land and the land's communities, that is understood as an acceptable cost of doing business.

It is hard to get the public and public leaders to see any issue of land abuse as urgent. Can that indifference be penetrated by a mere book written in love for what has been destroyed and for what remains?

Well, we had better hope so.

—Wendell Berry
Port Royal, Kentucky
November 2005

LOST MOUNTAIN

PART ONE

INTRODUCTION

Sometimes when sitting idly at my computer, I'll go to the federal Office of Surface Mining website and click on "Statistics." A line at the top of that page announces, "The number of tons of coal mined under the Surface Mining Law (since October 1, 1977) is now . . ." Right below, a ticker rolls constantly, tallying the 25-plus billion tons of coal that this country has produced and consumed in the last twenty-seven years. In the time it takes me to type this sentence, the number will jump from 25,366,669,740 to 25,366,669,940. That is to say, 100 tons of coal are extracted every two seconds in Kentucky, West Virginia, Wyoming, Pennsylvania, and a handful of other states. American coal companies extracted over one billion tons of coal in 2004, and 40 million tons more in 2004 than in the previous year. Nearly 70 percent of that coal comes from surface mines. Ninety percent of that fed coal-fired power plants to provide electricity to more than 50 percent of American homes.

This ticker has quickened its count over the last two decades because of a particularly destructive form of strip mining that has earned the name "mountaintop removal." Instead of excavating the contour of a ridge side, as strip miners did throughout the '60s and '70s, now entire mountaintops are blasted off, and almost everything that isn't coal is pushed down into the valleys below. As a result, the Environmental Protection Agency (EPA) estimates that at least seven hundred miles of healthy streams have been buried by mountaintop removal—some say the number is twice that—and hundreds more have been damaged. Blasting on the mine sites has cracked the house foundations of valley dwellers and polluted thousands of family wells. Creeks run orange with sulfuric acid and heavy metals. Wildlife populations have been summarily dispersed. Entire ecosystems have been dismantled.

And those ecosystems are the most diverse on the continent. What compounds the tragedy of mountaintop removal in central Appalachia is that this disappearing forest, the mixed mesophytic, is home to nearly eighty different species of trees. It is the rain forest of North America, and it is falling fast.

One reason this kind of environmental devastation receives so little notice is that it happens out of sight, up on top of mountains, where few people go. Unless you are flying over Perry County, Kentucky—half of whose mountains and forests have been literally blown away by explosives—you don't often see the damage.

Over the course of one year, from September 2003 to September 2004, I watched at close range as one mountain in Kentucky was destroyed so its coal could be extracted and sold to twenty-two other states and other countries. I visited the mountain at least once a month. I hiked over a hundred miles as I climbed to its summit over and over, then explored its flanks and descended along its headwater streams.

Today there is no summit. It may be too obvious an irony that this particular ridge was called Lost Mountain. But it is the truth, and now Lost Mountain exists only on topo maps of Perry County, Kentucky. The real thing is gone.

What follows is an account of events I witnessed over the course of that year. It is the story of how the richest ecosystem in North America is being destroyed, and how some of the poorest people in the United States are being made poorer by a coal industry that operates with little conscience or constraint.

THE NEW CANARY

It is, for my money, the most beautiful songbird in North America. The cerulean warbler bears a white breast, a thin black necklace, and a brilliant blue crown that slopes down its back into a deeper blue mantle streaked with black. The trick is ever getting a good look at one. Even by warbler standards, this is a small bird, and it nests in high canopy trees, deep within undisturbed Eastern broadleaf forests. I had only seen pictures.

So last spring, during this songbird's breeding season, I took the Tom T. Hall Highway to Buffalo Branch, a 3,600-acre woodlot in eastern Kentucky. From there, I pulled on hiking boots and followed Patricia Hartman, along with her two technicians, into the forest. Hartman is a young ornithologist who is conducting a three-year study of the cerulean warbler. She wants to know in which trees this songbird prefers to nest, and what kinds of vegetation surrounds those nest sites. What makes her work so pressing is that cerulean warbler populations across Appalachia are plunging—down 70 percent since 1966. And it is no coincidence that those forty years have also seen the most extensive destruction of Appalachian forests by strip mining.

Wearing Carhartt work pants and a Florida Gators T-shirt, Hartman led us through an underbrush of mayapple and spicebush. Redstarts and Acadian flycatchers sang from hidden perches. As I walked under a young maple, one of Hartman's technicians said, "Look up." Just above my head hung a tight little sack of a nest, dangling from just a few strands of grass looped over a thin branch. It was no bigger than a tennis ball, but it held the brood of a red-eyed vireo.

In the distance, we could hear three fluty notes, followed by a quick trill—the cerulean warbler. "There's not a lot of good hard information about them, because they're so hard to study," Hartman told me as we stepped over fallen branches. The cerulean warbler is a forest obligate—which is to say, it needs a large, undivided tract of woods to protect it both from predators and from parasitism by female cowbirds, who lay their eggs in the nests of other birds. And because the cerulean warbler nests so high, its diet and courtship rituals have largely eluded ornithologists. So far, Hartman had detected forty-five individual singers in Buffalo Branch and located three nests.

About a mile into the forest, Hartman stopped in a densely wooded glen and aimed her binoculars up the trunk of a chestnut oak. "Good, she's here," Hartman said, pointing to a forked branch near the top of the tree. I couldn't make out anything through my own binoculars, so Hartman handed me hers. Then the small brown nest came into focus. And jutting out from it were the female's white tail feathers, trimmed in black.

"Has she laid her eggs?" I asked.

"I think so," Hartman said. "Otherwise there would be more males hanging around, trying to get in on the action."

Meanwhile, Hartman's assistants had located the mate and were quickly tying orange Mylar ribbons to the trunk of every tree where he paused to sing. They timed his visit to each one, furiously registering the data in small spiral notebooks. Then they turned their binoculars back to the treetops.

"I've got him."

"Where?"

"The tulip poplar. Forty-five seconds."

"There he went."

"In the hickory?"

"Thirty feet up."

"I see him."

I kept aiming my binoculars in the same direction as theirs. But by the time I would find the right branch, the warbler would be gone. I listened again for his song: *zeep zeep zeep zizizizi zee!* Then I saw something dart about fifteen feet to another tree crown. I turned up my binoculars to another mass of leaves. As I tried to bring them into focus, a bright blue sliver pierced their green cover, flashing like a sapphire. The warbler made three quick hops along the branch, sang his quick run, and shot out of sight. It lasted only a few seconds, but I had seen him—this stunning and threatened passerine, this elusive forest singer who is quickly losing his summer breeding ground.

A more famous warbler, the canary, earned a reputation in Appalachia for the times it didn't sing. Miners once took caged canaries into underground shafts because these birds were especially sensitive to odorless methane gas, which leaked from coal. When too much methane accumulated in a mine shaft, the canary stopped singing. Far fewer cerulean warblers are now singing in Appalachia. Their silence, like the canary's, is also an indication of much larger problems.

September 2003

LOST MOUNTAIN

Look hard and you can find Lost Mountain in grid 71, coordinate B-10 of the *Kentucky Atlas and Gazetteer.* According to that topo map, the summit rises 1,847 feet above Lost Creek, whose headwaters come to life on the mountain's north face. This morning I left the bluegrass region of central Kentucky, where I live, and drove east along the Mountain Parkway, where the last of rolling grasslands, dotted with black tobacco barns, finally give way to the Cumberland Plateau, the foothills of what may be the oldest mountain range in the world—the Appalachians. From there, a narrow two-lane follows the meanderings of Lost Creek, so named because early hunters frequently lost their bearings when they ventured too far from the stream itself. In the 1920s, the midwife-turned-folk-singer-turned-union-activist Aunt Molly Jackson wrote a ballad that included this telling verse:

These Lost Creek miners
Claim they love their wives so dear
That they can't help giving them
A baby or two every year.

When the blacktop ends, I follow an old logging road that winds up and around Lost Mountain, ending at its peak. I set the parking brake on my truck and get out to take a look around.

Curiously, a fire tower that was standing a year ago has been blown or torn from its foundation and sent crashing down the ridge side. But even without the tower's perspective, I can see to the north thousands of acres—former summits—that have been flattened by mountaintop mining. Where there were once jagged, forested ridgelines, now there is only this series of plateaus—staggered gray shelves where exotic grass struggles to grow in crushed shale. When visitors to eastern Kentucky first see the effects of mountaintop removal, they often say the landscape now looks like the Southwest, a harsh tableland interrupted by steep mesas. I too have traveled through Arizona and New Mexico in the late spring, when ocotillo and Indian paintbrush are in bloom, and I understand the allure of that harsh landscape. But this is not the desert Southwest; it is an eastern broadleaf forest. At least it should be.

There was, of course, a time in this region when union miners would have extracted the coal with hand picks and shovels in deep underground mines. But twenty-five years after Jimmy Carter signed into law the Surface Mining Control and Reclamation Act (SMCRA), the coal industry has developed much more expedient and much more destructive methods of mining. I came to Lost Mountain because last month, Leslie Resources Inc. was granted a state permit to shave off its summit. I came to see up close what an eastern mountain looks like before, during, and after its transformation into a western desert.

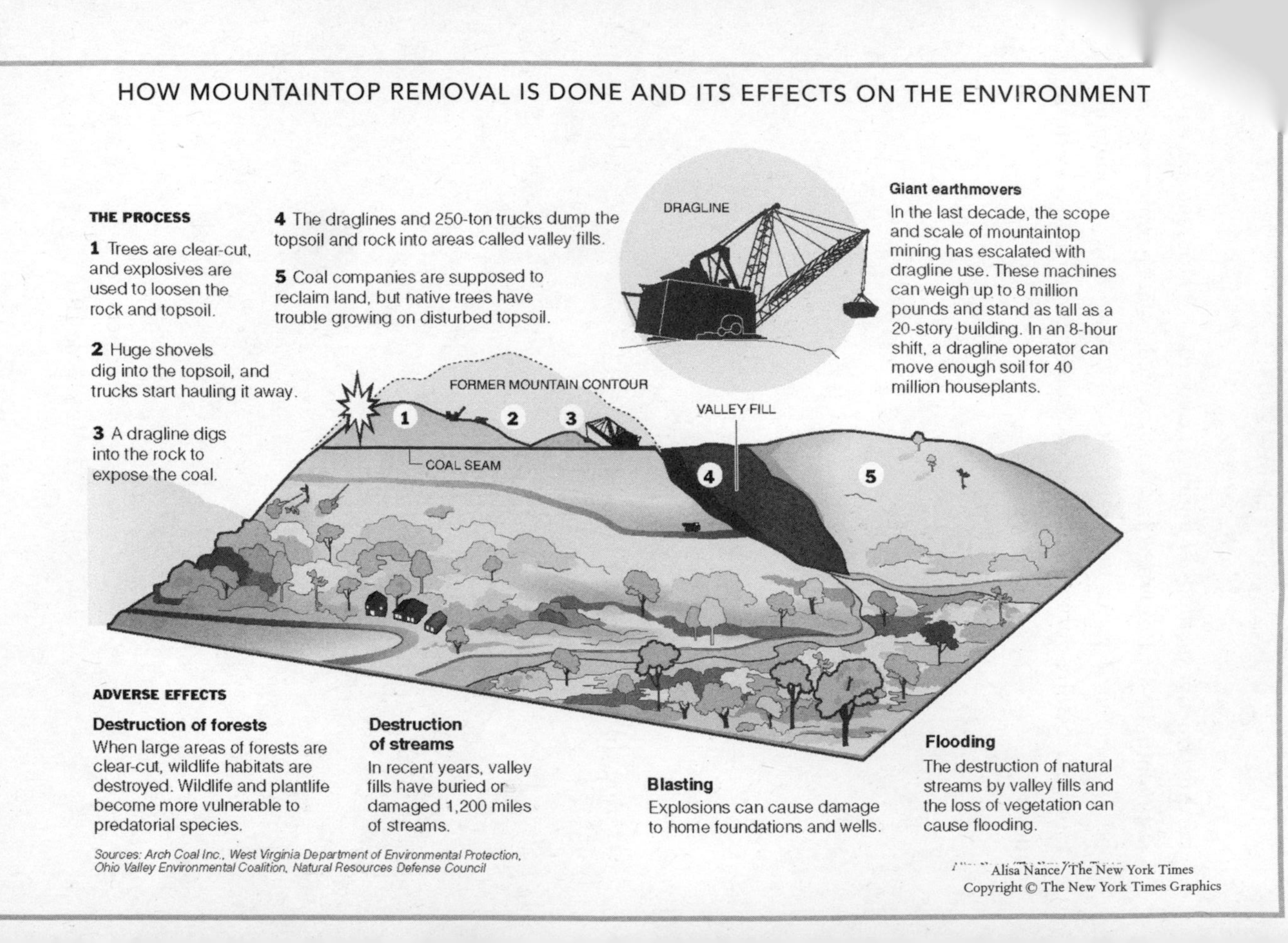
HOW MOUNTAINTOP REMOVAL IS DONE AND ITS EFFECTS ON THE ENVIRONMENT
THE PROCESS
1 Trees are clear-cut, and explosives are used to loosen the rock and topsoil.
2 Huge shovels dig into the topsoil, and trucks start hauling it away.
3 A dragline digs into the rock to expose the coal.
4 The draglines and 250-ton trucks dump the topsoil and rock into areas called valley fills.
5 Coal companies are supposed to reclaim land, but native trees have trouble growing on disturbed topsoil.
DRAGLINE
Giant earthmovers
In the last decade, the scope and scale of mountaintop mining has escalated with dragline use. These machines can weigh up to 8 million pounds and stand as tall as a 20-story building. In an 8-hour shift, a dragline operator can move enough soil for 40 million houseplants.
FORMER MOUNTAIN CONTOUR
VALLEY FILL
COAL SEAM
1
2
3
4
5
ADVERSE EFFECTS
Destruction of forests
When large areas of forests are clear-cut, wildlife habitats are destroyed. Wildlife and plantlife become more vulnerable to predatorial species.
Destruction of streams
In recent years, valley fills have buried or damaged 1,200 miles of streams.
Blasting
Explosions can cause damage to home foundations and wells.
Flooding
The destruction of natural streams by valley fills and the loss of vegetation can cause flooding.
Sources: Arch Coal Inc., West Virginia Department of Environmental Protection, Ohio Valley Environmental Coalition, Natural Resources Defense Council
Alisa Nance/The New York Times
Copyright © The New York Times Graphics

COAL: AN AUTOBIOGRAPHY

During the mid-'80s, I briefly worked at the same coal-fired power plant as my father in Louisville, Kentucky. He was a mechanical engineer, installing scrubbers inside the mill's three towering stacks. The scrubbers were designed to neutralize most of the sulfur and nitrogen oxide before these gases escaped into the air. I was a janitor. My job was to push a dry mop around the plant for eight hours, sweeping up all the soot that spewed from the subterranean furnaces. Every morning, a fine black dust again covered every inch of the plant; every morning, I put on a beat-up hard hat and pushed the mop up and down those long concrete landings. It was dreary work. I felt as though I were living inside a Springsteen song without the drums and guitars. There was just the plant: loud, hot, ugly, inescapable. All I had going for me, careerwise, was my improbable membership in the International Brotherhood of Electrical Workers—the IBEW. I paid dues; I went to meetings. I can't say I ever felt much brotherhood, but the union did guarantee me a decent wage, and I liked the *idea* of belonging to a union. My father wore a collared shirt and the shiny white hard hat that signified management, but I was a workingman.

Fortunately, that delusion passed and I went back to college. I thought I was through with coal. Working around that bituminous, burning rock was a nasty business; there had to be a better way to make a living. I wasn't thinking then about global warming or air quality, mountaintops or acid rain. I was just tired of that black film clinging to my skin.

While I was in school, my grandfather convinced one of his fishing buddies to sell me an old canoe for cheap. My grandfather was a forceful country preacher and people had a hard time telling him no. Years before, he had taught me how to balance the wind on the stem of a wood-and-canvas canoe with something called a J stroke. He had bought his Old Town canoe straight from the Maine factory in 1946. In that canoe, we navigated the coves and coastline that surrounded my grandparents' parsonage near Chesapeake Bay. So once I had a canoe of my own, I started exploring the rivers and forests of Kentucky. I paddled through the limestone palisades of the Kentucky River, the stretch of Elkhorn Creek that Walt Whitman mentions in *Leaves of Grass,* the Green River, where I once watched a bobcat drinking at the edge of Mammoth Cave. Truth told, I had already abandoned my grandfather's faith and had replaced his whitewashed sanctuary with the river's unroofed church and the forest's primitive icons—what Wes Jackson, president of the Land Institute, once called "altars of unhewn stone." It would have horrified him to know the hand he had played in my conversion to pantheism, but I could never understand how seeing the divine in the natural world constituted heresy, and I still can't.

A couple of summers ago, I spent one month exploring Robinson Forest, the largest contiguous forest in Kentucky. The state's purest streams flow from its watersheds, which are home to over sixty species of trees. Hiking and writing, sketching and wading, I conducted my days in various states of what E. O. Wilson has called "the naturalist's trance." I didn't have to read billboards or drive past miles of fast-food architecture. I didn't have to look at anything ugly or man-made.

I was going on in this vein one day to John Cox, a conservation biologist who was tracking the movement of an elk herd that summer. John did much of his work from a small prop plane, and he suggested I take a flyover with him to see things from that perspective.

I hadn't been on a plane since 9/11 and wasn't thrilled about the prospect. But the next morning, at a small airport near Hazard, Kentucky, I climbed into a three-seat Cessna with Cox and a young pilot named Neil. We lumbered down a narrow runway that stretched across a former mountaintop, then lurched into the air, rising above the canopy trees. A narrow blue thread, the North Fork of the Kentucky River, wound under ridgelines that together looked like the ocean's surface—an unfolding series of staggered green waves.

And then the color drained away, the trees dropped back. I suddenly was watching a black-and-white movie. All I could see below me was a long gray flatland, pocked with darker craters and black ponds filled with coal slurry. It wasn't just here and there—the desolation went on for miles. The tops of the mountains had been blasted away with the same mixture of ammonium nitrate and diesel fuel that Timothy McVeigh used to level the Murrah Building in Oklahoma

City. Dozers had carved up the rubble into this shifting landscape of vertical rock faces and long gray benches. A vast circuitry of haul roads wound through the rubble. It looked as if someone had tried to plot a highway system on the moon.

Farther ahead, something green finally came into view. Surrounded by miles of gray flats, a grassy knob stood strangely, surreally undisturbed.

"It's a cemetery," Cox said into our headsets. "The law says they have to stay one hundred feet from it on all sides."

The plane veered closer and then I saw it—the one-room chapel with its green metal roof; the white, rounded headstones; the tulip poplars that circled the edge of the knoll. This cemetery was the only sign of life in a deeply depressing landscape.

We were approaching a steep ledge, what in the parlance of the coal industry is called a "highwall." Here, one-half of the ridgetop had been blown away, leaving this vertical contour exposed. The plane tilted and Cox pointed to the bottom of the high wall.

"See that black line down there?" Underneath a hundred feet of bedrock ran a thin, barely visible seam of coal. "That's what it's all about," John said. "In twenty years, there won't be any mountains left here." There would be only these barren tablelands.

When we finally reached Robinson Forest, and were again flying over an unbroken ripple of green treetops, it felt like a great reprieve. Surrounded by such a sickly landscape, the forest looked so vital, so indomitable. But by that point I had already figured out that my grandfather's religion was wrong on another score: It doesn't take a whole lot of faith to move mountains; it takes about ten men and a company called Caterpillar.

We circled back toward the airport, back over the unnatural plains. Down below, I could barely make out a line of coal trucks creeping away along haul roads, heading for a coal tipple and then a power plant like the one where I used to sweep floors. I had quit coal back then at the stage of its conversion into energy, but now I found myself once more drawn back into its complicated hold over my home state—this time at the source of its extraction.

October 2003

LOST MOUNTAIN

Before the mining has started, I follow on foot the old logging road that winds up to the summit of Lost Mountain. There are, to my mind, two ways of thinking about a mountain: as something to be conquered, or as something to be revered. The conqueror is after personal gain in the sense of either taking something from the mountain, or scaling it in order to say "I did it." The conqueror, therefore, goes straight for the summit. By contrast, one who reveres a mountain either admires it from a distance, as in the paintings of Thomas Cole, or climbs it in a state of mind that is meditative, almost prayerful. Many Native American tribes refused to climb certain mountains, because they deemed those heights sacred. Though America's greatest mountaineer, John Muir, scaled the glaciers of California and Alaska, he never bragged about his climbs, but rather wrote about them with the ecstatic reverence of an acolyte. This is the same wilderness mindset

E. O. Wilson calls "the naturalist's trance," and I try to place myself in it as I slowly climb from the hemlocks up to the pines.

Four-wheelers command this dirt-and-gravel road by day, but four-legged mammals and members of the pheasant family take it back at night. At one muddy wheel rut, I stop to sketch the tracks of a deer, a fox, a raccoon, and a wild turkey. Then I drop down into the forest proper, the watershed that feeds Lost Creek. After extricating myself from a blackberry thicket, I climb noisily over a barricade of fallen tree limbs. A crow warns a white-tailed deer of my approach, and the doe hoofs it up over the ridgetop. When she is gone, I find myself standing beneath an austere canopy of tulip poplars. This is Kentucky's state tree, and it grows as straight as a flagpole. Daniel Boone once hollowed out a sixty-foot canoe from a single tulip poplar and packed his family down the Ohio River in it. The tulip tree is also the first hardwood to establish dominance after a deciduous forest has been cleared by fire, a blow-down, or in this case, chain saws that chewed through here about forty years ago. These poplars have already lost their leaves, and sunlight fills the understory of younger sassafras, hickory, and sugar maple. The woods are quiet except for a pileated woodpecker; the songbirds are already vacationing in Belize and other points south. Given time, one hundred years or so, oak, beech, and hickory would come to dominate this transitional forest. Three different communities of highly diverse trees would eventually agree on a silent charter about how best to inhabit these elevations. But that's not going to happen here.

I wander on down the ridge. Without thinking, I begin to follow the moist furrow of an intermittent stream. Such a lacework of tributaries feeds the lower creeks, but as the name implies, they only flow during wetter periods. I step around moss-covered cobble and maidenhair ferns that grow in the shape of delicate tiaras. Colonies of liverworts cover some of the rocks like small, green scallops. These modest-looking organisms actually carry on pretty fascinating sex lives. The liverworts needs moving water to spawn. And its preferred habitat seems to be these rain-catching, intermittent streams. During a downpour, the male liverwort extends a tiny, umbrella-shaped antenna. When a drop of rain hits it, sperm explodes inside that raindrop and bounces a couple of feet, where hopefully a female liverwort has sent up a little umbrella of her own to catch the flying spores.

In this way, the unassuming liverwort dramatizes one of the issues at the heart of mountaintop removal. In response to the charge that such mining methods bury hundreds of miles of central Appalachian streams, Bill Caylor, president of the Kentucky Coal Association, is quick to point out that an intermittent stream, such as this one, is not really a stream at all, because there are no fish in it. But consider the assumptions behind such a belief. According to this line of thought, if something, like the liverwort, is of no immediate and *obvious* use to *us*, then it is of no use at all. That modest flora like the liverwort are helping to hold rich soil in place, purifying water downstream, and providing habitat to other small animals like salamanders—or that they even hold an intrinsic value beyond what we might understand today—is a logic to which *Homo sapiens americanus* seems curiously immune. But we cannot bury entire ecosystems with bulldozers and then discover their value later. It will be too late. And as Aldo Leopold said in the middle of the last century, the first rule of the tinkerer is to save all the pieces:

> The outstanding scientific discovery of the twentieth century is not television, or radio, but rather the complexity of the land organism. Only those who know the most about it can appreciate how little is known about it. The last word in ignorance is the man who says of an animal or plant: "What good is it?" If the land mechanism as a whole is good, then every part is good, whether we understand it or not.

But of course we are no longer tinkering, we are leveling entire mountaintops. And as for where the pieces have gone, consider this sad commentary: When Neil Armstrong landed on the moon, some older eastern Kentuckians refused to believe it had happened. They were sure NASA had simply gone up to a strip mine at night and taken pictures of a guy in a space suit.

When I reach the mouth of the intermittent stream, I follow Lost Creek until I can see no signs of human intervention, not even the inevitable Bud Light can. I sit down on the bank, beneath the yellow glow of beech and maples. Dark water glistens in the shallows below. Squirrels rustle through the leaves. Trees decay where they have fallen, providing shelter and food. A Carolina wren hops among the tangled branches. These days, it is thought unfashionable, even backward, to talk about *laws of nature* or to read a philosophy, a *morality*, into the workings of the natural world. For 4,000 years, theologians and philosophers have debated whether an Intelligent Designer stands behind it all. I have nothing to contribute to that discussion. But this much seems clear: This *forest* certainly demonstrates an *intelligence*, one it has been honing for 290 million years. Its economy is a closed loop that transforms waste into food. In that alone, it is superior to our human economy, where the end of the line is not nutrients but toxic industrial waste. Is there *design* behind this natural intelligence? I have no idea. But I will venture this: *The forest knows what it's doing.*

Compare these two economies: the forest's and ours. The sulfur dioxide that escapes coal-burning plants is responsible for acid rain, smog, respiratory infections, asthma, and lung disease. In 2000, the Clean Air Task Force, commissioned by the EPA, determined that coal-fired power plants account for 30,000 deaths per year. In Kentucky, the number of children treated for asthma has risen almost 50 percent since 2000. Because of acid rain and acid mine runoff, there is so much mercury in Kentucky streams that any pregnant woman who eats fish from them risks causing serious, lifelong harm to the child she carries. A National Academy of Science report warned that 60,000 babies born in the United States each year could have been exposed to enough mercury in utero to cause poor academic performance later in life. Of the 113 tons of mercury produced each year in the United States, 48 tons comes from coal-fired power plants. Furthermore, as we all know but choose to ignore, fossil fuels—coal and oil—are responsible for the carbon dioxide that is making the planet hotter and its weather more volatile. This year, climatologists found record-high levels of CO_2 in the atmosphere. At the same time, forests worldwide have shrunk from 5 billion hectares (12 billion acres) at the beginning of the twentieth century to 2.9 billion hectares (7 billion acres) now; over 2,000 square miles of Appalachian forests will be eliminated over the next decade under current mining regulations. Because of such deforestation, 12 percent of the world's birds are endangered, as are 24 percent of its mammals and 30 percent of its fish.

A forest, by contrast, can store twenty times more carbon than cropland or pastures. Its leaf litter slows erosion and adds organic matter to the soil. Its dense vegetation stops flooding. Its headwater streams purify creeks below. A contiguous forest ensures species habitat and diversity. A forest, in short, does all the things that the mining and burning of coal cannot—that is its intelligence.

WHICH SIDE ARE YOU ON? (PART 1)

In 1998, a group of West Virginia environmentalists filed a federal lawsuit arguing that mountaintop mining leaches acids and heavy metals into nearby streams, and thus violates the Clean Water Act, which allows for the dumping of only clean "fill," not "waste." A year later, U.S. District Judge Charles H. Haden II sided with the environmentalists and handed down a decision that said debris from mountaintop mining was clearly waste, which could not be deposited in streams. Because Judge Haden's decision had the potential to shut down massive strip mine operations throughout the coalfields of Appalachia, the Clinton administration moved quickly to settle the West Virginia case. It promised to create an environmental impact statement (EIS) that would reexamine the effects of mountaintop removal and rein in coal operators.

Work on the statement slowed considerably when the Bush administration took office. In the meantime, the Department of the Interior, under the direction of former coal lobbyist Stephen Griles, rewrote a key provision of the Clean Water Act, which states that "fill material" can be deposited in American waterways, but "waste" cannot. Griles reclassificed all waste associated with strip mining as merely benign "fill material." Judge Haden rejected that change as well, countering that "only the United States Congress can rewrite the Act to allow fills with no purpose or use but the deposit of waste." But in January 2003, the conservative U.S. Fourth Circuit Court—on which Supreme Court Chief Justice Roberts was serving at the time—overturned Haden's decision. Finally, in May of 2003, the Draft Programmatic Environmental Impact Statement was released, and its findings, along with its proposals, have occasioned the latest round of arguments about coal and conservation.

The 5,000-page document admitted that mountaintop removal is bad, and for the usual reasons: it buries headwater streams, causes erosion and flooding, degrades water quality downstream, kills a lot of aquatic life, shakes the walls and cracks the foundations of nearby homes, and wipes away huge portions of an extremely diverse ecosystem. The solution? The National Environmental Policy Act (NEPA) states that any EIS must offer alternative possibilities to current practices. Furthermore, the explicit purpose of the mountaintop-removal EIS, as stated by the EPA, was to "minimize, to the maximum extent practicable, the adverse environmental effects to waters of the United States and to fish and wildlife resources from mountaintop mining." Griles ignored all this. Instead his "alternative" version of the EIS proposed streamlining the mine-permitting process and doing away with the rule that requires a hundred-foot buffer zone between streams and mine sites. Coal operators could still fill up to two hundred and fifty acres of a watershed with the rubble that was once a mountaintop. Mine sites could still leach toxic acids into creeks where small valley communities once performed baptisms.

It was these conclusions that brought activists and counteractivists pouring into downtown Lexington, Kentucky, one summer afternoon in 2003. At the bottom of Broadway, about a hundred protesters had gathered. You've seen them before. They wear tie-dye shirts, Birkenstock sandals, and earnest expressions. They give fiery speeches about corporate greed and human arrogance. They play bongos. Two blocks up Broadway, in front of the Kentucky Coal Association headquarters, a group of middle-aged white men, thick around the middle, were having a cookout and waging a counterprotest. You've seen them too—the golf shirts, the khaki shorts. They hang out in cigar bars and like the president. At the bottom of the hill, the signs said things like "Solar and Hydro Power to the People." Up the street, one placard read "If It Can't Be GROWN It Must Be MINED."

The real show, however, was unfolding on top of the covered pedway that extended over Broadway and, in effect, divided the two factions. A guy in a rappelling harness and a hard hat had scaled the hotel wall on one side of the pedway and shimmied out on top of it. From there he dropped down two huge banners on either side. One announced "King Coal Is Killing Kentucky," the other "Stop Mountaintop Removal." Traffic started to back up. Police gathered down below. A fire truck arrived and raised an extension ladder to retrieve the troublemaker. I stood with the neutral crowd that had gathered around the spectacle. From there, I could see both the environmentalists and the coal men, and that's when I remembered a protest song I hadn't thought about in years, "Which Side Are You On?" Florence Reece wrote it in 1931 during eastern Kentucky's brutal union wars. Since then, many union drives have used the song as a rallying call, changing the words to fit the particular fight. But as I watched the police handcuff the young radical and pack him into the squad car, I realized the sides had changed. Now the fight is no longer between underground union miners and the big bosses, but between the surface-mining industry and the people who would like to see mountaintops stay where they are.

November 2003

LOST MOUNTAIN

This morning I cast my vote for the Kentucky gubernatorial candidate who accepted the fewest contributions from the coal industry and who promised, if not to revoke strip-mining laws in this state, at least to enforce them. Now, driving up the muddy switchbacks of Lost Mountain, I can see a thin column of gray smoke rising over the next ridge. As I round a bend near the summit, the forest falls away below my driver's side window. I am not speaking figuratively. The trees that lined the left side of this road two weeks ago, and that held in place the southern slope of Lost Mountain, are gone. Stumps line the road. Down below, all the ground cover and topsoil had been churned under—"grubbed"—by D-11 bulldozers. Nothing but mud, rock, and fallen trees remain.

I park my truck out of sight of the workers below and sit down on one of the stumps. *Scalped* is the word that keeps repeating itself in my notebook: This mountainside had been scalped. The trees that covered it now lie in massive piles all down the slope. The pile at the bottom has become a burning pyre, and a haze of smoke fills this concave southern valley. One dozer, with its long crescent blade, is slowly pushing the other piles down into the fire.

It seems to me like arrogance compounded by ignorance. While a sustainable, value-added timber industry is this region's most promising economic alternative to coal, this coal company, along with many others, doesn't even bother to save the timber it cuts. By contrast, my fellow Kentuckian Wendell Berry, who has visited the tribal forests of the Menominee Indians in northern Wisconsin, reports that when the Menominee were forced to give up hunting and gathering in the nineteenth century, they turned to logging and forest management. In 1854, their 220,000-acre forest was estimated to hold 1.5 billion board feet of wood. Today, after 140 years of sustainable logging, the figure remains the same. The Menominee example is a model my state might take seriously. But here on Lost Mountain, these fallen trees are instead being shoved onto a burning heap.

On the next ridge over, another dozer is pushing boulders out of the way to carve a haul road for the coal trucks. All around me there is nothing but rock, smoke, and ravaged soil. Then I see something that makes this scene even sadder. Standing a few feet away is a single green seedling, shooting out a dozen small branches. Somehow the dozer missed it. And now, the entire emptiness of the slope gathers around this seedling like an unbearable presence, a ghost forest.

And not just any forest. What heightens the tragedy of surface mining in central Appalachia is that the chain saws and dozers are stripping away the oldest and most diverse forests in North America. A million years ago, the Pleistocene glaciers forced northern trees to slowly migrate south. But the glaciers never reached this part of the Appalachians. And when the massive ice sheets finally retreated, they left in their wake a landscape that looked much like a modern strip mine. Consequently, over hundreds of thousands of years, the Appalachians were responsible for reforesting most of North America. But no forest ever achieved its diversity of tree species. It remains the continent's seedbed, its mother lode.

In 1934, a formidable botanist named Lucy Braun bought her first car and, along with her sister Annette, spent the next twenty-five years studying the broadleaf forests of North America. Both sisters earned their Ph.D.s from the University of Cincinnati, and they lived and worked together all their lives, never marrying. (Once, when a friend asked Lucy to go with her to a dance, Lucy replied, "How can you let strange men put their arms around you?") Though Lucy was the younger sibling, Annette always deferred to her, as did most people in the emerging field of forest ecology. The Braun sisters logged over 65,000 miles in their car as they ranged across the eastern United States. Lucy came to love the central Appalachian forests most of all. She was not intimidated by legends of moonshiners and mountain feuds. She was a small woman who showed the mountain people a courtesy that was not always forthcoming from outsiders. And here, on the Cumberland Plateau, she catalogued over eighty different species of trees. The name Braun gave this forest community was *mixed mesophytic*, because it was a middle climate—not too hot or too cold, too wet or too dry. Unlike most forests, Braun found, the mixed mesophytic had an astonishing number of mature trees making up its canopy; no species dominates. It is because of Braun's discoveries that today many naturalists refer to these small mountains as the "rain forest of North America."

What had stood here two weeks ago was not a mixed mesophytic forest, not yet. But it would have become one. Instead, the industrial equivalent of an ice age glacier will soon scour Lost Mountain, and the hopes of any kind of forest will be gone.

Back at home three hours later, I turn on the election results. My candidate has been soundly beaten.

THE POWER TO MOVE MOUNTAINS

There is no better place to understand the semiology of a strip mine than at Goose Pond, a five-acre "reclamation" area that sits in the middle of a massive mountaintop-removal project called the Starfire Mine in Breathitt County. One might begin, literally, at the top, with the term *overburden*. What is burdened in this case is the coal seam down below; overburden is the oak-pine forest, the topsoil, and the two hundred feet of sandstone that stand between the coal operator and the coal seam. Of course, to whom this xeric forest is a burden depends largely on who profits from extracting this coal and who pays for living down below the mine site. When the overburden is dislodged, it becomes *spoil*. And as we have already seen, according to the 2002 rule change, the spoil that is dumped into the valleys below is no longer waste, but "fill." Streams are not buried; rather, valleys are filled.

Reclamation, however, is the term that finally puts us squarely in the realm of Orwellian slipknots. Speaking plainly, to reclaim something is to get it back. The 1977 Surface Mining Control and Reclamation Act (SMCRA) requires coal operators to restore the "approximate original contour" (AOC) of the land they have mined. According to the Kentucky Department of Natural Resources, "The condition of the land after the mining process must be equal to or better than the pre-mining conditions." Scanning the reclaimed portions of the Starfire Mine site, I can see hundreds of acres of rolling savanna, planted with lespedeza, an exotic legume imported from Asia, one of the few plants that will survive in this shale. But in what sense does a savanna "approximate" a summit? In what sense is a grassland monoculture "equal to or better than" a mixed mesophytic forest? The reality is that mountains pitched at a grade as steep as that of the Appalachians cannot be restored. Gravity and topography are working too forcefully against you.

Perhaps sensing that the AOC stipulation would be a hard sell, lawmakers added this provision to SMCRA: Coal operators could obtain an "AOC variance" if they could prove the post-mined land would be put to "higher or better uses." In the beginning, that meant commercial or residential development. A few housing complexes and even a prison were built on these sites (when the prison watchtowers started to lean because of subsidence, locals dubbed the facility Sink Sink). But there weren't nearly enough developers clamoring to fill these barren flats with strip malls or apartment complexes. And anyway, it was much cheaper to plant grass on an abandoned mine site and call it a "pasture."

That is what has happened at Goose Pond, a five-acre pond surrounded by fescue and crown vetch. Rock islands dot the still water and hardscrabble locust trees cling to the cobble. Above the pond stands a small wooden observation deck, replete with two of those mounted viewing stations like those you find at Niagara Falls. Looking only through these binoculars, one might indeed be fooled into thinking that this is some kind of wildlife sanctuary. A couple of swans are even gliding around the pond. But back away two steps and you will see a behemoth blue crane sitting in a deep pit behind the pond, swinging back and forth like the fin of a mechanical shark. Attached to the crane is a dragline that rakes a house-sized maw across the coal seam, scooping up a hundred tons of rock at a time. Down inside the pit sits an explosives truck. Its tank reads, THE POWER TO MOVE MOUNTAINS. Coal trucks rumble past on all sides of the pond. Highwalls frame the horizon. But none of that matters. A coal company has only to erect the flimsy trappings of a tourist stop and they have converted a wasteland into a "public use." Who wouldn't want to fish for trout in the shadow of a dragline?

Still, my favorite part of Goose Pond is a large sign that stands next to the observation deck. It is covered with pictures of pink lady's slipper, large-leaf magnolia, ruffed grouse, painted trillium, spotted salamander, and a wood rat. It reads, "These are some of the other wildlife and plant species you may encounter during your visit." The word *other* adds a particular absurdity to what is already an outrageous lie. Wandering the forest below this mine, I have been lucky enough to see each of the flora and fauna pictured on this sign. But one is about as likely to find ruffed grouse or painted trillium on Wall Street as to find them up here on these hundreds of deserted acres. All the species pictured are inhabitants of a deciduous forest, the likes of which this place won't be able to sustain for another thousand years.

CAT

December 2003

LOST MOUNTAIN

I have heard numerous stories about do-gooders and documentary filmmakers who set out to inspect a strip mine only to find themselves confronted with the blade of a D-9 dozer, raised to windshield level, and bearing down fast upon them. I decided not to show my wife the article in this morning's paper about a state surface-mine inspector who died after being found beaten and unconscious in his home, his body mutilated by human bite marks. But I also can't quite shake that image, so I've decided to follow a less conspicuous road that leads around the backside of Lost Mountain, through a small community called Harveytown.

Christmas lights outline the trailers and small clapboard houses clustered tightly at the creekside, as if gravity itself had set them there. This western side of the mountain shows fewer scars from logging. A series of sandstone outcrops stretches along the ridgeline, providing vital shelter to smaller animals, including the endangered wood rat. This unfortunately named rodent is actually quite handsome, with large eyes and long whiskers. Americans particularly should feel a certain affinity for this rapacious collector of baubles. If something shines—a piece of glass, aluminum foil, a shotgun shell—the wood rat takes it home. He piles the loot just outside his nest, which sits back in a narrow rock crevice. No one knows why. Perhaps he just wants a bigger midden pile than the wood rat living in the next rock house.

All through Appalachia, wood rats are on the decline. A parasitic roundworm has decimated much of the population. The northern wood rats seem to be moving south and setting up colonies here in Kentucky. But it might not be the smartest move, since the "edge effect" caused by strip mining—the creation of smaller woodlots with an increased circumference—has made it much easier for foxes and bobcats to prey on them.

Earlier in the fall, I climbed along this same ridgeline, going from one rock formation to another, looking for evidence of wood rats. My search was hampered by the fact that many leaves had already fallen, covering the middens. Still I went on raking through them until I picked up one clump of brown leaves and found that I was holding a copperhead. Now, while it's true these mountains harbor a fair number of snake handlers, I for one have never been sympathetic to their cause. I say anyone who tempts a notoriously angry god by tossing around venomous reptiles has got it coming. Nor did this particular copperhead seem to care much for his part in the drama. Fortunately, the weather had been cold, and he was listless. I jumped backward, quickly unhanding the serpent. We each collected ourselves and abandoned the boulder. If there had been any wood rats there, the copperhead had surely taken care of them.

Now, three months later, I'm climbing this ridge again, this time to get a decent look at the mining on the other side without drawing any attention. I can already hear large equipment churning. At the ridgetop, I circle north and crouch under a stand of young trees. Two enormous backhoes are clawing away at the substrate directly beneath me. Slowly they cut a vertical rock wall all along the inside of this hollow to lay the wide road that coal trucks will need in order to maneuver this mountainside. Along with bulldozers and graders, they have already cut a quarter-mile road, stretching down to the highway. Eventually, the road will reach the top of the mountain, providing access to the top three coal seams that lie beneath Lost Mountain. But here, at mid-elevation, the work has gotten tougher. The dozers are no longer shoving aside topsoil, as down below; now they are struggling with boulders. They carve away as much of an opening as they can while the backhoes load the loosened rock onto haul trucks. When those huge beds are full, the trucks begin creeping up a makeshift path to a flat bench near the top of the mountain. I watch through binoculars as they back to the edge of the bench. Slowly, a hydraulic lift raises the bed, sending eighty-five tons of brown sandstone and gray slate spilling down the mountainside. Then, like Sisyphus with a driver's license, each one repeats the whole laborious process again and again.

"WAS IT ALL BY DESIGN?"

Coal operators are not an easily intimidated bunch. But there is probably no one in the state of Kentucky who rattles their cage like a forty-eight-year-old grandmother named Teri Blanton. A prime mover in Kentuckians for the Commonwealth (KFTC), the state's largest social-justice organization, Blanton has spent the last two decades helping coalfield residents fight the corporations that have turned so much of eastern Kentucky into what she calls a toxic dump. A few months back, Blanton was going through some old family photos with her daughter and a niece she is now raising. In one picture, Blanton is holding her scowling two-year-old niece. When the girl, now thirteen, asked why she had such a sour look on her face, Blanton's daughter Jennifer replied, "You were probably on a picket line somewhere." As it turns out, she was.

One can get a real education in environmental corruption and smashmouth class warfare by tracking the past twenty years of Blanton's life. She grew up in a small town called Dayhoit, in Harlan County, where four generations of her family had lived along White Star Hollow. It was the kind of community where neighbors shared their coal in the winter, and on a rare piece of flat land, one man, Millard Sutton, grew enough vegetables to feed nearly everyone in Dayhoit. Families took turns helping out in his garden. Blanton moved to Michigan in the '70s to start a family, then moved back to Dayhoit in 1981 as a single mother of two. Her career as an activist started shortly afterward when she phoned the highway department and asked for someone to clean up the large puddle of black water and coal sludge that stood in front of her trailer where her children caught the school bus. The highway department called the coal company that was mining around White Star Hollow, and the company responded by sending a coal truck to slowly circle Blanton's trailer all day. "That really burnt my ass," Blanton recalled, "that they thought they could shut me up by intimidation." That coal company, owned by two brothers, James and Aubra Dean, never did clean up the mess, and in the end, after Blanton's relentless badgering, the highway department built a private road up to her trailer.

Unfortunately, Blanton's problems were about to get much bigger than a slurry puddle. Since moving back to Dayhoit, Blanton's two children had been constantly sick. Sometimes, after bathing, they would break out in what their doctor called "a measles-like rash." But they didn't have the measles. The groundwater that fed their well had been poisoned with vinyl chloride, trichloroethylene, and a dozen other "volatile organic contaminants," or VOCs. On a three-acre plot a half-mile from Blanton's home, the McGraw-Edison Company was rebuilding mining equipment. In the process, they sprayed trichloroethylene-based degreasing solvents on transformers and capacitors. They piped PCB-laden transformer oil directly into Millard Sutton's large garden. They even sprayed it on the dirt roads of the next-door trailer park to, as they said, "help keep the dust down." They were just being good neighbors.

In the late '80s, Blanton, along with two other Dayhoit women, Joan Robinett and Monetta Gross, began pushing the EPA to test their water. "In the media, we were portrayed as these hysterical housewives who didn't know what we were talking about," Blanton recalled. Finally, after many of the wells were found to be contaminated by chemicals from the plant (which had since been sold to Cooper Industries), the EPA declared Dayhoit a Superfund site in 1992 and put it on the National Priorities List of hot spots. "I moved back to Harlan County thinking I was bringing my children home to a safe place," Blanton said. "Instead I brought them back to a chemical wasteland."

The EPA excavated five thousand tons of contaminated soil from around the plant, then trucked it to Alabama, where it was stored next to a poor African-American community. To extract the contaminated groundwater, a pump-and-treat system was installed on the site of the abandoned plant. This catalytic oxidation unit filtered out some of the VOCs and released the remaining elements, including carcinogens, into the air to be quickly dispersed. At least that was the plan. Blanton began researching pump-and-treat systems and found that they had only been tested at two airports, where winds were high. But White Star Hollow was a different story. "It's like a bowl where the fog sits down on the river until the middle of the day," Blanton said. "And I lived right at the fog line—and as the crow flies, not very far from the plant. In my mind, I knew they were going to poison me and my kids all over again."

On a bright, cold day in November, I drove with Teri Blanton back to Dayhoit. We passed the trailer park that sat next to the McGraw-Edison plant, as well as the field that used to be Millard Sutton's garden. Blanton pointed to the yellow house next to it. "Everyone in that house died of cancer," she said. And she said it more than once as we passed houses, traveling up the hollow.

The road followed Ewing Creek, running brown from recent rains. And then, as we followed it farther upstream, the water turned orange. Blanton pointed for me to pull off at a rusting cattle gate, where a sign read MOUNTAIN SPUR COAL COMPANY. We got out and climbed the gravel road that led to an abandoned strip mine. A nasty orange syrup called acid mine water was pouring out of a pipe that drained an open mine pit. The sulfuric acid collected in a small pond, then spilled over into the creek below. Blanton lit a cigarette. "I grew up on this creek," she said. "I grew up walking these mountains and I've watched them crumble before my very eyes. It just makes me angrier and angrier knowing that these people can operate in such a manner and get by with it."

For years, the Dean brothers, along with a third partner, Carl McAfee, have been playing an elaborate shell game that keeps them in business and free from any responsibility to the land or local landowners. It works like this: The three men own one company that remains in good standing with state regulators. Then they set up smaller companies with names like Limousine Coal, Master Blend, and Mountain

Spur. These operations lease equipment from the "good" company, and post a small bond that will supposedly cover the cost of reclamation should the company declare bankruptcy. Which is exactly what they do. The shell company forfeits its bond, which is never enough to complete the reclamation, and local communities are left with cracked foundations, a contaminated creek, poisoned wells, and steep slopes that pour mud down when it rains because there is no vegetation to hold the soil in place.

Last year, Blanton tried to block any further permits from being issued to these shell companies. Before a hearing officer from the Office of Surface Mining, she laid out an extensive paper trail showing at least one of the men, or their wives, was named as an "incorporator" or an officer in every one of the companies that had abandoned reclamation and declared bankruptcy. But curiously, the OSM ruled that it could find no clear link between the companies. And the fight to penalize violators of SMCRA was made even harder by a puzzling 1999 verdict in the U.S. Court of Appeals for the District of Columbia. There, in *National Mining Association v. Department of Interior*, the court ruled that permits could not be denied to companies for violations at mines they no longer controlled. So all coal operators like the Deans have to do is declare bankruptcy, start a new company, and move on to the next permit.

Across the creek from where we stood, I could see the home of Blanton's childhood friend Debbie Williams. Before the mining started here, she spent $7,000 to have a well dug. But as soon as the blasting began, her faucet was running as orange as the water now draining from this mine. The foundation of her house and her chimney have been cracked. The Deans have not yet been forced to reimburse Williams, and it's likely they won't be. But just up the ridge, another one of their companies, Sandlick Coal, continues to strip away the trees and the coal.

Before leaving Dayhoit, Blanton and I stopped at the White Star Cemetery, which sits up in a small clearing. Some of the headstones were so old I could barely tell them from the large rocks that had rolled down the mountainside. "Hey, this is pretty," Blanton said. "I don't think I've ever been up here on a day I wasn't burying someone." Many of the newer tombs were set aboveground in cement block vaults. Blanton pulled back some plastic flowers beside one of her cousins' markers. "She lived next to what we called the killer well," Blanton said. "Everyone who lived around that well died."

In the middle of the cemetery were buried two of Garrett Howard's three sons, the two that were born after he started working at the McGraw-Edison plant. "They both developed non-Hodgkin's lymphoma before they were thirty and died," Blanton said. We stared in silence at the dates on the markers. "Almost nobody in Dayhoit lives past fifty-five," she went on. "At the meetings, the people from the EPA would accuse us of being too emotional. I told them, 'Let all of your family members and friends die around you and see if you don't get emotional.'" She knelt beside the grave of a high school friend. On the headstone was a depiction of a father and son standing beside a stream. "He was a real redneck," Blanton said, breaking into a smile. "I loved him."

As we drove out of Dayhoit, past Dorcus Jane Lane (named in honor of a reliable bootlegger), I realized the most sinister part of this whole sad story is that it was all done intentionally. A multinational corporation hid in a hollow of one of the poorest counties of one of the poorest states, and knowingly dumped hundreds of deadly chemicals right on the ground. Then, to add insult to irony, a Virginia coal company picked up where it left off, this time flooding out its residents, poisoning their wells, and killing their creek.

But for Blanton, the web of control reaches further than that. "We were fueling the whole United States with coal," she said of the region's last hundred years. "And yet our pay was lousy, our education was lousy, and they destroyed our environment. As long as you have a polluted community, no other industry is going to locate there. Did they keep us uneducated because it was easier to control us then? Did they keep other industries out because then they can keep our wages low? Was it all by design?"

One might dismiss such questions as the kernel of a conspiracy theory. But the death of Blanton's father from black lung and of her brother from a mine collapse are facts. The deaths of Garrett Howard's young sons from lymphoma are facts. The White Star Cemetery is a collection of terminal facts.

January 2004
LOST MOUNTAIN

Forty years ago, Lyndon Johnson first visited eastern Kentucky, the poorest place in America, to declare his War on Poverty. Ever since, Appalachia has been a cause for liberal Democrats with presidential aspirations, right up to the late Paul Wellstone. I asked Greg Howard, director of the grassroots art organization Appalshop, why this was. "Hollows look enough like suburbs to northerners," he told me. "Poor white children look enough like their own kids."

In the spring of '63, Harry Caudill published his stark portrait of the region, *Night Comes to the Cumberlands*. The *New York Times* sent Homer Bigart down to see if it was really as bad as Caudill said. Bigart reported that it might be worse. In October, he published a series of front-page articles that described children so hungry they ate dried mud chinking from the sides of houses. He told of miners who were paid a dollar an hour to perform the most dangerous work in America. John F. Kennedy read the articles and summoned Kentucky governor Bert Combs to the White House. Combs confirmed Bigart's assessment of the region, and Kennedy pledged $45 million from his executive funds to provide winter relief for the poorest Kentuckians. Furthermore, he told Combs that after an upcoming trip to Texas, he would fly down to Kentucky and see things for himself. But a bullet prevented Kennedy's visit, and so the poverty tour fell to Lyndon Johnson.

Not that Johnson didn't have some populist credentials. He had begun his political career as a New Deal Democrat. And as Caudill wrote, "This was the only time in history when a President of the United States came to visit a portion of the American people distinguished only for their chronic, deep destitution." Johnson's helicopter landed in far-eastern Martin County, where the unemployment rate stood at 70 percent. His limousine maneuvered the pocked roads, and with Lady Bird at his side, Johnson stopped at some tar-paper shacks to assure a few families that he would bring them into his Great Society.

Forty years ago, Appalachia's poverty rate stood at 31 percent. Since then, nearly 2,300 miles of roads have been laid across the region and 800,000 more families have indoor plumbing. And today, eastern Kentucky's poverty rate hovers around, well, 31 percent. Furthermore, one can look at a map of central Appalachia, and almost to the county—in Kentucky, Virginia, West Virginia, and Tennessee—the areas that the Appalachian Regional Commission deems "distressed" are the ones that have seen the most strip mining.

Back at Lost Mountain, I set my truck in four-wheel drive and slog up the ridge side and onto the muddy mine site, where a light snow is falling. Work has slowed as temperatures have dropped. The dozers have erased almost all the logging roads, shoving aside topsoil and subsoil in search of coal seams. In the wet soil, I can still make out wild turkey footprints inside the wider tracks of the bulldozers. If an apocalyptic deluge were to sweep down right now, these incongruous images would make for an interesting fossil—a clue, perhaps, for some future anthropologist as to what might have caused the last great extinction.

Perhaps I am being overly dramatic. Perhaps I have become too used to seeing the glass half-empty. Hanging around strip mines will do that to you. But this much is true: The twentieth century added more people to the world than all other centuries combined. Scientists calculate that each of the more than 6 billion humans on the planet needs 2.5 arable acres to produce the food and energy they need. The World Wildlife Federation estimates that the planet cannot regenerate its resources if every human being uses over 4.45 acres. Currently, the global average is 5.44 acres, which is still a deceiving number given that each American uses 23.47 acres. For the rest of the world to live like Americans, we would need four more Planet Earths.

In his essay "The Last Americans," Jared Diamond argued that while wealth and conspicuous consumption are certainly signs of social status, they may not indicate success for a society as a whole. In fact, if Mayan civilization is an indicator, material prosperity, overpopulation, resource consumption, and waste production are actually signs of a society's impending collapse. Like the coal operators of Appalachia, the Mayans stripped their forests and polluted their streams with silt and acids. Their population spiked steeply in the fifth century.

More and more people started fighting over fewer resources. By the time Cortés marched through the Yucatán, a culture that once numbered in the millions was gone. "Why," Diamond asks, "did the kings and nobles not recognize and solve these problems? A major reason was that their attention was evidently focused on the short-term concerns of enriching themselves, waging wars, erecting monuments, competing with one another, and extracting enough food from the peasants to support all those activities." It sounds to me like an unnerving assessment of this century's first four years.

It is natural, however, for human beings to think in the short term, according to E. O. Wilson. It's a habit we picked up in Paleolithic times: Those who worked for short-term gains lived longer and had larger families. And we Americans are masters of short-term thinking. We think in election cycles and the weeks between car payments. For a nation whose economy is based on planned obsolescence and ever-increasing consumer spending, thinking about the personal and the environmental future might hurt sales. But we have become too technologically advanced as a species to still think in terms of Paleolithic self-interest. Wilson warns that the technological advances of the twentieth century have led us to a bottleneck—a time and a place that is overpopulated and depleted of resources—and now only long-term planning will lead us out. Of the last century, Wilson remarks, "We and the rest of life cannot afford another hundred years like that."

Down along Lost Mountain's southernmost ridgeline, I nearly step in one of the holes that have been augered to determine the depth of the coal seam. The haul road has almost reached this particular peak. I walk along its razorback for what I know will be the last time.

Driving home, I pull off at the Coalfields Industrial Park just outside Hazard. I veer left past Trus Joist, a wood-products plant that received over $100 million in subsidies to locate in Hazard. I pass a coal tipple and a large slurry pond where a culvert coughs black sludge into the pool of black water. From there, the road winds on across five hundred acres of a former mountaintop-removal job. It is a cratered and rolling landscape like all the rest, sparsely populated by short-leaf pines and black locust. Obviously no attempt has been made to restore this former mountain to its "approximate original contour." Instead, this industrial park was touted as a model for the "higher and better" uses to which flat land could be put. Jobs, after all, were on the way. This industrial park was created with over $21 million collected by the coal severance tax. The state levies a 4.5 percent tax on each ton of mined coal. But that money goes into a general fund, and less than half of it is returned to the coal-producing counties that need it most.

I pass an aluminum-sided warehouse that some company has already abandoned. A sign advertises 4,100 square feet for rent. And then there is nothing. Just acres of brown lespedeza. If I didn't know where I was, I might guess I was driving across a flat Kansas wheat field. I keep going until the road dead-ends beside a square brick building and an empty parking lot. At the upper-left corner of the building, red letters spell out SYKES. Tampa-based Sykes Industries, a computer call center, moved into this industrial park in 1999. Then governor Paul Patton crowed that he had lured a "first-class company" to eastern

Kentucky, and that it would give the region's youth a reason to stay in the mountains. Put aside for a moment the question of how making less than $7 an hour with limited benefits would inspire a new generation of eastern Kentuckians to stay home. The point is that Sykes was offered an incentives package well into the millions of dollars, with no obligation to remain after those tax breaks expired. And the company didn't. In July 2003, just months after the tax abatement expired, Sykes pulled the plug on 393 jobs housed in this building and shifted them to El Salvador.

Indeed, it seems as if the city of Tampa has it in for eastern Kentucky. TECO, Tampa Electric Company, is one of the most negligent mountaintop removal outfits in Appalachia. The blasting and flooding caused by its mining have ruined hundreds of houses and displaced families. And here at the Coalfields Industrial Park, Sykes has callously used this poor region as a temporary holding station on its way out of the country to an even poorer city, where it can pay workers even less.

My neighbor back in Lexington, the late author and McArthur Fellow Guy Davenport, once wrote, "Distance negates responsibility." What do the citizens of Tampa know, or possibly care, about Sykes's cynical manipulation of 393 people in eastern Kentucky? What would they think of the way TECO has made valley communities almost uninhabitable throughout the coalfields?

If we follow Davenport's proposition out to its conclusion, we arrive, of course, at where we are—in a globalized economy where responsibility does not exist. Pakistani children sew together Nike soccer balls for 6 cents an hour. Immigrant workers hack off fingers in Midwestern meatpacking plants because fast disassembly lines mean cheap hamburgers. A Florida electric company exploits another region's people, extracts its resources, and takes the profits elsewhere. Debra Burke, a poor woman in McRoberts, committed suicide after TECO blasting ruined her home and flooding caused by the mine repeatedly wiped out her family's subsistence garden. There's blood everywhere you look.

WHICH SIDE ARE YOU ON? (PART 2)

After the Environmental Impact Statement on mountaintop removal was released, several forums were held in Kentucky and West Virginia to allow for public comment on the study. This struck me as a rather cynical formality, but I drove down to the hearing at the Hal Rogers Center in Hazard, Kentucky, to hear what coalfield citizens had to say. There were about 150 people in the auditorium, mostly men. They wore Carhartts and work boots; some still had on their hard hats. Onstage, the drafters of the EIS document sat at a long table. All the heavy hitters were there: the Environmental Protection Agency, the Army Corps of Engineers, the Department of the Interior's Office of Surface Mining, the Department of Fish and Wildlife. They sat attentive, with pens poised, ready to take the public pulse.

Anyone who wished would have five minutes to speak his or her piece. The first speaker, Bill Caylor, president of the Kentucky Coal Association, began by complaining for thirty seconds that he had prepared a twelve-minute speech and why couldn't somebody do something about this. Then Caylor, a man with thick white hair and a neat white mustache, asked for a show of hands of those who had come to support strip mining. If anyone besides me didn't raise a hand, it was hard to tell. Having sized up his audience, Caylor launched into a barrage of statistics: 120 million tons of coal were mined in Kentucky last year, placing the state third in extraction behind West Virginia and Wyoming; that coal fetches $3 billion annually; 80 percent of Kentucky coal is sold out of state; in the last fifteen years, coal-related employment has dropped 60 percent; at 4.1 cents per kilowatt-hour, Kentucky coal is the cheapest energy source around. "We're like the Saudi Arabia of America," Caylor announced in conclusion, meaning, I think, that Appalachia has a lot of fossil fuel, and not that the region's poor have been greatly oppressed by a wealthy minority that control said fuel.

Still I puzzled over his fact sheet. Coal jobs have dropped 60 percent precisely because strip mining requires far fewer men to operate much larger machinery. It seemed hardly an argument for *decreasing* regulation on mountaintop removal. And in what sense was it a good thing that 80 percent of Kentucky's coal is sold out of state? Why should Dayton and Detroit, or China for that matter, get the coal but be held accountable for none of the environmental consequences of its extraction? And if everyone is doing so well, why is eastern Kentucky the nation's capital for OxyContin abuse?

Caylor was followed by a long line of miners and mining engineers. One man read an exhaustive list of every Hazard business that had been built on flattened land. He ended with, "If that's not economic development, I don't know what is. And these couldn't have been built on the side of a hill." The crowd was generous with its applause. Several speakers drew attention to the deer, elk, and turkey that had returned to the region after mountaintop removal began. The wife of a miner pleaded for her husband's job, then asked, "What use are the mountains to us other than coal?" A young man who said he had just moved to the region spoke in bold terms: "I like mountaintop removal. I'd like to live on top of a mountain. Without mountaintop removal, I wonder what this area would look like." I was starting to wonder, too. Another man asked of the environmentalists in the room—a largely rhetorical flourish, as it turned out—if they would ride home on a horse or were going to sit on a block of ice to cool down when they got there. I took his point. I was going to do neither (and I had certainly consumed my share of oil writing this book). Everyone who spoke endorsed the EIS "alternative" to let the Office of Surface Mining regulate operations as it sees fit.

Then up stepped a gray-haired man in a blue suit and a flamboyant purple tie. He said his name was Paul David Taulbee. His grandfather had helped log the mountains back in 1912, and had then worked as a deep miner from 1915 to 1952. His father worked in the mines for twenty-eight years. I suspect there were country preachers in the family as well, because it quickly became clear that Taulbee had come to deliver a sermon. He wasn't talking to the men and women up on stage; he was addressing the congregation.

He was tired—sick and tired—of outsiders coming into eastern Kentucky and telling its people what they *should* do. He hinted at a conspiracy afoot by the rest of the state to keep eastern Kentucky poor.

He wanted the federal government and the Army Corps of Engineers and the Environmental Protection Agency to leave this region and its people *alone* to mine as they saw fit. "We want to be unfettered to develop to the fullest extent," he stormed, one finger held high. Were that to happen, all of those native people who had followed the outward migration north for better jobs would come back home. Eastern Kentucky would finally thrive. Because, Taulbee intoned, what the outsiders should never forget is this, "The only way to *stay* in the mountains is to *mine* the mountains!" A standing ovation followed. The moderator called for a short break. Everyone went out to smoke.

It was gray and drizzling. I started up my fossil-fuel-burning truck and headed home. I replayed the hearing in my head. It's true that the contour mining of the '70s cut out shelves in these mountains and made room for the chain stores that followed an expanded highway. But anyone who has ever looked down on the strip jobs from a plane knows there is enough flat land in eastern Kentucky to plop down ten thousand Wal-Marts. And much of that land is so inaccessible that a retail business or a housing developer wouldn't be interested.

The idea that this land is prime real estate is one of the industry's most popular arguments and one of its weakest. Topo maps show that there are now enough flattened mountains in eastern Kentucky to set down the cities of Louisville and Lexington. Yet developers are not exactly rushing to these plateaus where topsoil and water are scarce. Rather, the coal industry has created the ultimate supply-side economy, where it's hard to tell the difference between "real estate" and abandoned land. And if more than half a billion acres of mined land sit abandoned and unreclaimed, in what sense is it real estate at all? Mountaintop removal has robbed it of any intrinsic, ecological value, and its desolate state has robbed it of any market value. To call it real estate is simply corporate double-talk.

There is one exception—the golf course in Stumbo, Kentucky, built on a reclaimed mine site. Many in the coal industry like to praise its greens and fairways, but it seems to me like a particularly cruel joke in a region where most people do not belong to the socioeconomic class that grows up playing golf.

As far as the return of game animals, deer populations have risen all over the eastern United States, not just around abandoned mines. The elk that do graze around the edges of reclaimed sites were reintroduced a few years ago from western states. Strip mining had nothing to do with it.

As for Paul David Taulbee, I could tell him that surface mining accounts for only four thousand jobs in all thirty counties of eastern Kentucky, averaging out to 130 jobs per county. I could tell him the old deep-mining jobs aren't coming back, and the people who left for Cincinnati and Cleveland might not want to either, especially if they were coming back to wasted mountains and dead streams. I could tell him that if coal hadn't brought prosperity to the mountains in the last ninety years, it probably isn't likely to do so.

But Taulbee isn't going to listen to me. I'm an *outsider,* as he had said, the worst kind of elitist. The state flag for Kentucky reads "United We Stand," but it is a divide-and-conquer tactic that the coal industry likes to employ to pit eastern Kentucky against the rest of the state, especially its urban areas (never mind that most of the coal companies' headquarters are in other states). Only those from eastern Kentucky can speak for the region, or criticize the region, according to this argument. Any other voice is illegitimate; everyone else is an outsider. I once raised the notion of my outsider status with Teri Blanton, who grew up in Harlan County. "You're not an outsider," she replied. "We all live downstream." And she's right. While I would never presume to speak as, or for, someone from Appalachia, ecologically there *is* no outside. We're all in this together. And if something is wrong—as polluting a community's drinking water is wrong—how can it matter who makes that known? And if Kentucky is a commonwealth, then these mountains should be held in trust for all Kentuckians. More than that, they should be held in trust for future Kentuckians, who will need clean streams and jobs that do not rely on a finite resource. The miner's wife had asked, "When are you going to start thinking about us instead of the environment?" But perhaps the harder questions are these: When are we going to start thinking about both at once? How do we move beyond the which-side-are-you-on way of thinking and start giving careful consideration to *both* jobs and trees, people and streams?

I'm no economist, but I don't think it requires one to recognize the economic injustice of removing the natural wealth of a region without giving anything back to it. In addition, the years I have spent thinking about these mountains have led me to this conclusion: One of the first steps toward solving some of the economic problems of Appalachia lies in rejecting the idea that coal is "cheap energy," thereby forcing operators to pay the *true* cost of extracting it. The reality of our modern economy is that we attach no monetary penalty to *throughputs*, the toxic by-products and environmental damage that result from industrial manufacturing. But because we have settled for a linear, throughput economy where the by-product of energy is waste, that waste must be taxed. There must be a *cost* for polluting streams and rivers with mercury and choking them with sediment; there must be a *cost* for pumping sulfur dioxide and carbon dioxide into the air. Because natural capital such as coal is a limited resource, it must be taxed as such. In other words, market prices must reflect social and environmental costs. To have an economy based solely on the short-term growth of our gross domestic product follows a dangerous and absurd logic—that we can have infinite growth based on the use of finite resources.

A carbon tax levied on the use of fossil fuels would reflect this true cost and would discourage wasteful use. Kentucky, a midsized state, ranks eighth in the country in energy use. Why? Because we can afford to; coal is cheap here. Conversely, as of last November, Toyota dealers could not keep the fuel-efficient Prius in stock because oil prices were rising. By the same logic, when consumers are forced to pay the true cost of coal, they will begin to think about smaller homes, better insulation, fluorescent lighting, strategically placed shade trees, and solar hot-water heaters. The technology is there; we simply lack the will.

In addition, I believe a large percentage of the carbon tax should be returned to the coalfields in the forms of subsidies for jobs in reforestation—jobs that would lead to carbon sequestration and jobs that would begin to give real credence to the term *reclamation*. If a carbon tax was implemented to supply the materials and labor to reforest even 50 percent of abandoned strip mines with sustainable hardwoods, the unemployment rate would drop to zero almost overnight. Urban centers across the country should remember this: Central Appalachia is poor because so much has been taken from it and so little has been returned. The mountains of Appalachia are responsible for the illumination and air-conditioning of billions of houses, and neither the people nor the land has been properly compensated.

To create an industry around real reclamation in the mountains would be the first step toward turning the linear economy into a closed-loop economy that emulates the principles of ecology and sustains itself.* And to tax coal at its real cost in terms of what economists call "externalities" would, perhaps, hasten its demise and give way to clean, renewable energy. The coal industry has left central Appalachia a scarred and toxic landscape. It is time—and given rising temperatures and melting glaciers, there isn't much time—to revive the land and the people's quality of life.

---

*If all of that sounds too easy, too abstract, too far down the road, then consider something very tangible, like the coal severance tax. Currently in Kentucky, the tax on extracting coal is 4.5 percent for each ton, the same rate as 1976, before the destructive forces of mountaintop removal. West Virginia's severance tax is 5 percent and Wyoming's is 7 percent. In addition, over half of Kentucky's tax, 58 percent, goes into a general fund to be spent all across Kentucky. It is simply unfair that parts of the state that suffer none of the consequences of strip mining should receive money from the coal severance tax, money that often funds pet projects for large contributors to various election campaigns. In eastern Kentucky, some of the severance tax is given right back to coal companies in the form of subsidies. And $70 million more has gone to the eight industrial parks, with very meager returns. Eight and a half million dollars was spent on the Bluegrass Crossing Regional Industrial Park; today, one company sits there and employs thirty-five people making air bags. Unless each of those employees is making $245,000 a year, and pouring it all back into the community, that was not money well spent.

What to do? For starters, the Kentucky state legislature should raise Kentucky's rate to at least that of West Virginia, and preferably that of Wyoming. The price for a ton of coal has doubled in the last year; the industry can afford a higher severance tax. Secondly, all that money should be returned to the coal-producing counties, which are the poorest counties

---

in the state and some of the poorest in the country. And finally, the money must be directed into new local and sustainable value-added economies. Joining the global market will only make eastern Kentucky poorer, just as it makes Pakistan and El Salvador poorer. The best alternative is the opposite of globalism: a regional economy. Mark Dabenstott, a rural development expert with the Federal Reserve Bank of Kansas City, recommends three basic strategies for pulling regions like eastern Kentucky out of cycles of poverty and economic exploitation: thinking and acting regionally, finding an economic niche, and encouraging entrepreneurship. The first idea would help protect the region's resources and keep dollars circulating longer within local communities. The second plan would capitalize on the region's cultural strengths to find new markets; there is, for example, a strong tradition of beautifully crafted musical instruments and clean-lined Shaker furniture in Kentucky. If timber was sustainably harvested by local people, made into furniture by local people, and sold both locally and nationally to those who respect such craftsmanship, then eastern Kentucky might finally rid itself of coal operators who take most of their profits out of the state. And as for entrepreneurism, perhaps it is time to give local people, rather than multinationals, the subsidies that are needed to create a locally owned economy made up of furniture makers and cabinetmakers, tree farmers, fish farmers, foresters, and people raising nontimber forest products, such as mushrooms and herbs. The combination of locally owned businesses and a federally subsidized reforestation industry would go a long way to solving eastern Kentucky's poverty, its pollution, its flooding, its mudslides, its drug abuse.

February 2004

LOST MOUNTAIN

Leslie Resources has now blocked off every dirt road leading up Lost Mountain with imposing iron gates. I don't take it personally. Eastern Kentuckians are as attached to their ATVs as urban Kentuckians are to their SUVs, and while both do their respective share of environmental damage, it's the former that Leslie is trying to keep off Lost Mountain. NO TRESPASSING signs hang on all the gates.

If history is any indication, strip mine operators along Lost Creek may have reason to worry. On August 8, 1967, the *Hazard Herald* reported that $300,000 worth of mining equipment had been sabotaged at Lost Creek. One and a half tons of carbo-nitrate was used to disengage an auger, a D-9 bulldozer, two trucks, and two drills. A detective concluded that the vandals took the explosives from the company magazine on-site. "It had to be a professional job which required several hours preparation," estimated one official. "It definitely wasn't any kids."

By 1967, nearly every other coal-producing state in the country had laws protecting property owners from coal companies that own the rights to the minerals below. But in Kentucky, anyone could find their land literally stripped out from under them. There were no regulations on how the coal was to be extracted, and so the companies used the cheapest method: blasting away everything that stood between them and the coal. Retaliation seemed the only recourse that citizens had. The violence peaked in 1967, when another local paper, the *Mountain Eagle,* carried in a single issue five reward offers from five different coal companies, all seeking information about industrial sabotage.

Today coal companies must obtain permission from landowners before they can begin mining. And because the companies must also pay the landowners a small percentage of the profits, that permission is often given. To many eastern Kentuckians, strip mining is no longer seen as an act of political injustice but as a burden of economic necessity.

Since Leslie only leases the mineral rights on Lost Mountain, however, I tell myself it has no jurisdiction to keep me off property it doesn't own. I duck under one gate on the eastern side of the mountain and start walking. Chain saws have mowed down most of the trees on this slope, and they all lie where they fell. Only the dead trees have been left standing, and pileated woodpeckers move back and forth between them as if they can't believe their luck—nothing now stands between them and the carpenter ants that colonize diseased beech trees.

Higher up, where the hardwood trees still stand, I pass a sign tacked to one that reads DANGER BLASTING. Almost on cue, a siren sounds to signal a coming blast. I am still too far from the site to actually see the explosion, but two minutes later, when the blast sounds out over the hollow, I feel a slight trembling beneath my boots. After a few more minutes, a yellow plume is moving through the trees, carrying with it a sharp, sulfuric smell.

I drop down the ridge side to a lower logging road that leads directly to the source of the smoke. I can hear the constant beeping of haul trucks inching back to the edge of the hollow fill. A few months ago, I could follow this gravel road up to the mountaintop. Now I find it blocked by a row of boulders. Peering over them, all I can see is the top of a truck as it raises its bed and sends another load of rubble down into the valley. I can tell, however, that what used to be a ridgeline leading west is now nowhere in sight.

I circle back around the mountain and begin climbing through the younger trees and wild roses that still cover the northern slope. The last obstacles are the capstones that mark the summit. I shimmy into a narrow crevice of rock, find a foothold, and haul myself up. At the crest, doubled over and gasping, I still see in the dirt the same traces of wild turkey, grouse, and raccoons that I saw months ago. The mountaintop is still here, still as it was. These obdurate boulders attest to it. It's not until I reach the other side of this summit and look down that I see what has changed.

The lower ridgeline is nearly gone. What was, last month, a gradual slope leading westward is now, right below me, a fifty-foot vertical drop that gives way to dark pits and gray ledges.

You can think of this mountain, or any mountain in Appalachia, as a geological layer cake with seams of coal two to fifteen feet thick, separated by much thicker bands of sandstone, slate, and shale. The seams are numbered in descending order: The one nearest the summit is the Hazard 12 seam, and about three hundred feet below lies Hazard 9.

The narrator of Merle Travis's famous folk song "Sixteen Tons" begins his lament with:

I was born one mornin' when the sun didn't shine
I picked up my shovel and I walked to the mine
I loaded sixteen tons of number 9 coal
And the straw-boss said, "Well, bless my soul."

That same number 9 coal seam lies beneath Lost Mountain, but no deep miners are trying to dig it out. Why bother? Why send hundreds of miners burrowing underground when a few men armed with explosives and bulldozers can blast right down to the seam? And whereas in the '40s it took one miner all day to load sixteen tons of coal out of a deep mine, today one man behind the wheel of a loader can, in five minutes, fill a coal truck with sixty tons of this bituminous rock. What makes strip mining so cost-efficient is precisely what makes it so devastating.

Here on Lost Mountain, the crew goes straight for the highest three seams, where there is less earth to move and a readier supply of coal. The dozers have pushed much of the vegetation and topsoil to the edge of this man-made plateau, called an "area mine." The twisted trees and mounded dirt form a berm around the darker crater. Young maples and hickories stubbornly hold on at the edge of the mining, where so much of the topsoil has been upturned and compacted. What compounds the problems of mountaintop removal is that when the bedrock is disturbed, it increases in mass by 20 percent; that additional matter is called "swell" and will eventually be dumped down into the valley below.

Staying out of sight, I loop down to the edge of the mining and duck in behind three toppled pine trees. From here I have the whole scene in front of me. At the far edge of the mine site, a white "powder tower" now stands, filled with explosive material. In front of it, dozers have shaved down to the number 10 seam. A loader scrapes the coal into mounds, then shovels them into the first coal trucks to climb Lost Mountain. Those trucks will take their loads five miles up Highway 80 to Leslie Resources' coal tipple, which sits beside the North Fork of the Kentucky River. There the coal will be processed and loaded into railcars.

Closer to where I'm crouched, dozers push the loosened rock into piles, and front-end loaders fill one haul truck after another with the debris. The trucks dump their load down the mountainside and return for more. It's all extremely efficient. Like ants in a colony, everyone keeps moving, diligent in his single task.

I can almost make out the tattoo on the forearm of the man operating the loader right below where I'm crouched, and I wonder what it must be like to work in such a place day after day? "The interior landscape," nature writer Barry Lopez tells us, "responds to the character and subtlety of an exterior landscape." But in a place that has neither character nor subtlety, what does a man's interior landscape start to look like?

The whistle blows at 4:30—quitting time. The workers grab their lunch coolers and jump down from the dozers, trucks, and loaders. I retrace my path down the backside of the mountain. My face and arms bear the scars of blackberry gauntlets, and my water bottle is empty. My thoughts have turned from the ravages of strip mining to the shelves of cold beer at the BP station down below. I am not looking ahead, not looking at anything really, when the huge silver maw of a bulldozer comes lunging over the ridge about twenty feet in front of me. The driver doesn't see me; he is cocked at too steep an angle. I leap back over several fallen trees and take cover. Whatever else bulldozers do, they do not move fast. This one backs down the hill, coughs another cloud of black smoke into the air, then lurches back into view, shoving topsoil to the side. The driver pauses each time to get his bearings, and each time I get another look at the huge, serrated blade. For the first time I understand completely why Harry Caudill described it as a "monstrous scimitar."

Once the driver has cleared a space to work, he sets about the real task—knocking down trees. I'm startled to see how easily a twenty-year-old maple succumbs to the dozer's blade. The dozer is graceless and resolute. Each time the driver backs down the hill to take a run at another tree, I scramble about fifty yards farther away. When I am finally far enough down the mountain to escape the driver's notice, I take a seat on a stump. It is almost dusk, and the mountain has darkened to a silhouette. I can no longer see the dozer. But from the stump, I watch as one tree after another falls against the violet light of the setting sun.

ON BAD CREEK

Daymon Morgan reached down to pull up a white flower by its roots. We were standing on a hillside at the head of his hollow on Bad Creek, near Harlan. He brushed away the soil to show me a thick red rhizome. "Bloodroot," he said. "The Indians used it for war paint." With his pocketknife, Morgan cut open the rhizome and brushed three orange strokes across the back of his hand. Then he tucked the roots back under a layer of humus.

It was early April. "Not too much is up yet," he explained. "But when they are, this hillside is covered with herbs: black cohosh, trillium, goldenseal. Food, medicine—I was raised on them." And he was raised right here, on these steep slopes. As a young man, he harvested ginseng in these hills. Ginseng, thought by many Asians to have life-prolonging properties, can fetch as much as $400 for one root in China. Consequently, most of it has long been poached. Now Morgan usually gathers cherry birch, peppermint, and sassafras for an herbal tea. Before we started up the mountain, he poured me a glass. It was delicious.

I first met Daymon Morgan at one of the public hearings in Hazard where both sides were sounding off about the Bush administration's proposal to allow mining within a hundred feet of streams, a practice prohibited by the Clean Water Act. Several engineers from TECO Energy had been extolling the job growth that coal brought to the region. Then Morgan stepped to the podium, wearing denim overalls, a red flannel shirt, and a white straw cowboy hat. He looked to be in his seventies. "These people who talk about coal bringing jobs," he began, "why, you wouldn't have no problem at all selling them a sky hook." I wasn't sure what a sky hook was, but I was pretty sure this was the genuine article—a self-made, freethinking mountaineer, stepping right out of the past.

The engineers from TECO had also claimed that many headwater streams were actually dirtier before mining, when local people dumped their garbage in the hollows. To that Morgan replied, "You're right. They're not as dirty now. Because they're not there." He said the stream that ran beside his family's cabin is now buried under sixty feet of what used to be a mountaintop called Huckleberry Ridge. And all around him, Leslie Resources continues leveling mountains and burying streams. After the hearing, I asked if I could come down to take a look.

Sure enough, when I reached Morgan's modest log cabin that April morning, he and his wife were sitting on their porch, listening to the incessant beeping of haul trucks creeping along the ridgetops all around them.

"Let's take a ride," Morgan said, and we climbed onto his six-wheel amphibious ATV, called the Argo. He was already well tanned and he moved like a much younger man. He wore the same overalls, flannel shirt, and straw hat that he had had on at the hearing, and I realized this was a kind of uniform that lent him a yeoman's authority to speak for the place where he grew up.

Morgan was drafted right out of high school and sent to Okinawa and Iwo Jima. "When I was in the service, I thought a lot about this land," he said, "how I used to hunt on it as a kid, and how I used to come up here and sing. I decided to save up my money and buy this place when I got back." And he did. He paid $1,000 for a hundred acres, which today is the only stretch along Bad Creek that hasn't been strip-mined.

We followed gravel switchbacks up the mountain to the spot where Morgan pulled up the bloodroot. He pointed out a rare white chestnut tree. Down this ridgeside, Morgan has set out a row of chestnut saplings. He imported them from Virginia and hopes they will survive the blight that wiped out the native chestnuts in the '40s. Back then, as the oldest of twelve children, Morgan worked with his father on the family's scratch farm. "By the time we were old enough to carry a hoe, we had to work," Morgan said. "We'd take a hillside like this and grub it up and plant corn and beans. That's the way we made a living. We didn't need no fertilizer, the soil was so rich. We had forty or fifty head of hogs and we turned them back up in here." Many other families did the same. The hogs would feast all summer on beechnuts. Each mountain farmer had his own brand for his hogs, kept on record at the county courthouse. Morgan's father's brand was a split in each ear. "When he was feeding them, my dad would take a sharp knife and split their ears," he recalled. "They didn't even quit eating. Just shook their heads a little bit."

It was a marginal economy for marginal land, but it sustained a family of fourteen. Soon, though, the railroads that had been laid alongside the Cumberland and Kentucky Rivers would signal the end of such sustenance living. An extractive economy had arrived. Once the major timber barons were through, there were no beech trees left and no beechnuts for the hogs.

Morgan cranked up the Argo and we ventured higher. He pointed to an outcrop hidden by taller trees. "The man I bought this land from," Morgan shouted over the 18-horsepower engine, "that's where he set his still. His wife had a bell down at the house and she'd ring it if any revenuers came around." We rode across the ridgeline that marks the boundary of Morgan's property. Everything from here down to his cabin belonged to him; everything over the ridge—at least the mineral rights—belongs to one coal company or another.

We rode past the last stands of chestnut oaks and mountain laurel. Then we made a hard right and were suddenly driving across the savanna landscape so typical of a postreclamation mountaintop-removal job. Brown lespedeza waved like prairie grass. "This was reclaimed over thirty years ago," Morgan said, but it had only a few pines and scattered black locusts to show for it. When we reached the edge of this shelf, Morgan shut off the Argo. Across the valley stood another scene I was becoming used to. The same company that was mining Lost Mountain had reduced one more forested watershed to another dark wasteland.

"That used to be a big beautiful mountain," Morgan said. "Now look at that."

A black butte rose up from a series of staggered black plateaus that stretched out against the horizon. Narrow rivulets caused by erosion ran down the sides of the benches. The characteristic gray haul roads wound through it all. The ugly panorama dwarfed the line of dozers that sat below the nearest bench. And the only thing that keeps those dozers off Morgan's property is a piece of legislation he spent twenty years fighting to enact.

In the 1880s and '90s, a native Kentucky schoolteacher named John C. C. Mayo began riding through the eastern counties on horseback, offering gold dollars to farmers who would sell him their mineral rights. Since the farmers made their living off the surface of the land, selling what was underneath seemed a good idea. Many signed a contract called the "broad form deed," so named because it gave the deed holders broad rights to extract the coal by any means they desired. The farmers obviously imagined that miners would tunnel under their land using picks and shovels, then haul the coal out with ponies. At the end of the nineteenth century, no one who sold their mineral rights could have imagined the industrial evolution that would lead to strip mining.

Mayo bought up thousands of mineral parcels, which he then sold or leased to coal companies. The companies themselves soon got into the game of buying mineral rights under the broad form deed. Near the end of the nineteenth century, Kentucky River Land and Coal Co. paid as little as a quarter an acre for the coal under Morgan's property. And when he bought these hundred acres back in the '40s, the company still had every right to strip off every pound of topsoil to claim their coal.

This happened all over eastern Kentucky. Men who had learned to drive tanks during World War II had no problem climbing onto D-9 bulldozers and cutting benches along the side of a mountain. Then, in 1961, the Tennessee Valley Authority, a major provider of hydroelectric power, decided to get into the coal business. TVA signed contracts to buy 16.5 million tons of strip-mined coal.

Under the broad form deed, the mining was ruthless and the landowners powerless. Mrs. Bige Ritchie, who lived on Sassafras Creek, watched a bulldozer plow through a family graveyard. It upended the coffin of her infant son and pushed it over into the creek. "I like to lose my mind over it," she told Ben A. Franklin of the *New York Times*.

Things finally came to a head in 1965, thanks to a sixty-year-old widow and an eighty-year-old coffin maker. When officials from Caperton Coal Co. told Ollie Combs that they intended to strip-mine her property, she quite sensibly replied that her land was too steep and the mining would bury her house. But on returning from the post office one day, the woman who became known as the Widow Combs found the bulldozers moving in. A small woman dressed in a head shawl and overalls, the Widow Combs sat down on a boulder in front of one dozer's menacing blades and refused to move. Eventually, the Knott County sheriff and a deputy arrived. They picked the Widow Combs up by her armpits and feet and dragged her off her own property. By then, however, a Louisville *Courier-Journal* photographer, Bill Strode, had arrived on the scene, and he caught that image on film. The sheriff arrested him too, and both Strode and the Widow Combs spent Thanksgiving of 1965 in jail. But by then, it was too late. The next day, Strode's photograph appeared on the front page of most major newspapers, and that image of the Widow Combs seemed to stir the public conscience as nothing had before.

Later that year, several bulldozers pulled onto a stretch of land along Clear Creek Valley, where an eighty-year-old coffin maker, known to his neighbors as Uncle Dan Gibson, lived with his second wife. The land belonged to Gibson's stepson, who was serving with the Marines in Vietnam. When the dozers crossed onto the property, Gibson met them with a .22-caliber rifle, which he cocked and set across his lap. He told the operators that the man who owned the land was serving "over the waters," but that Gibson himself would personally see to its preservation. "I got seven shells in this gun," Gibson told the workmen, "and I'll lead you if you don't go."

They went. But a few hours later, Kentucky River Coal officials, along with six sheriff's deputies, returned. One Seargent Mitchell charged Gibson with "breach of the peace," but Gibson replied that he wasn't answering to "anybody who'd root the dead out of the ground for a block of coal." When the deputies moved in, Gibson warned them, "The only person who's going to take me out of here is the undertaker." The deputies moved back.

Finally, late that night, an agreement was reached. Gibson said he would put his gun down if the coal company promised not to return. The sheriff took Gibson to jail, but his neighbors surrounded the property, bearing .22s and shotguns, until Gibson was released.

For a region in sore need of heroes, the Widow Combs and Uncle Dan Gibson became instant icons. Citizen groups started forming to fight the broad form deed. Daymon Morgan joined just about all of them and served as chairman of the Citizens Coal Council. And as a member of Kentuckians for the Commonwealth (KFTC), he discovered that a powerful point could be made by exposing regulators to two particularly uncomfortable emotions—embarrassment and shame. When Harry Snyder, the head of the federal Office of Surface Mining, reduced a company's fine from $112,000 to $5,000, Morgan and some other members of KFTC came up with a plan. Snyder was due to speak at a state mining directors' conference, being held at the Stratford House in Fredericksburg, Virginia. From a building across the street, Morgan scoped out the hotel. "I saw there were two sets of doors," he remembered. "So half of us hit one set, and half of us hit the other. There were all of these good-looking girls serving drinks to everybody. This one little woman went up there and grabbed the microphone from Snyder's hand and told him what a crook he was." Morgan leaned back in the Argo's seat, crossed his arms, and smiled. "They didn't call hotel security or nothing. They just sat there in shock. So we all got up there and said what we had to say. Then finally we thanked them for inviting us to their gala, and we left."

Finally, in 1988, Kentuckians voted for a state constitutional amendment that required coal operators to get a landowner's permission before mining. That law saved Morgan's home and land, but he still has to look out every day at the thousands of acres that have been destroyed all around him.

"I belong to a lot of peaceful organizations," he said as we stared down at the lifeless mine site. "We believe in dialogue. Well, I believe in dialogue, too. But sometimes . . ." His voice trailed off. He was choosing his words carefully. "We're fighting terrorism right now," he began. He wasn't talking about Islamic militants; he was talking about American strip miners. "If people are going to poison you to death, I think we should do whatever is necessary to put a stop to it. The state and federal government won't do nothing. I don't want to say take the law into your own hands. That's a big step. But . . . I don't know."

Morgan turned around in his seat to take in an as yet unscathed range just to the left of the strip mine. Color was coming back to the red maples. Cherry and serviceberry trees were blooming. "See how that holler zigzags in and out of those mountains?" Morgan said. "That's where Main Creek runs down into Greasy Creek. Sometimes I think I'll just take my gun and my dog, go walking down through there and just keep going." For several moments, we watched his other, imaginary self disappear down into the valley, beyond the strip job, back into the past.

IR
The Power To Move Mountains
1198B

March 2004

LOST MOUNTAIN

It's 80 degrees and sunny. I'm driving fast alongside Lost Creek, with the windows down and Neil Young wailing on an old cassette. Redbud and white dogwood are blooming along the steep slopes and up green gorges. Down below, junked cars and hot-water heaters are rusting on some grassless patches of land in front of deteriorating trailers. What is remarkable about the ugliness of Appalachian poverty is its closeness and contrast to the spectacular mountains rising around it.

If the day were not so nice, I might be chastened by the number of wooden crosses that each mark a spot where someone met a violent death. Take one of the poorest parts of the country, add to it alcohol and pills, hard curves and coal trucks, and what you get is a lot of little white crosses staggered along the roadside.

I set my parking break at the usual spot below Lost Mountain and hike to the top. On the bench right below me, a white truck carrying a large tank pulls into view. To get a better look, I shuffle about thirty feet down the ridgeside and wedge myself between two boulders that I hope will provide decent cover when the blasting starts. *Flyrock* is the rather benign term for everything that scatters when the explosives are detonated. Regulators from the state Department of Natural Resources tell me they have cracked down on blasting violations in the last few years. People were getting hurt, chimneys and house foundations were cracking, dishes were shattering. An older couple in Knox County was sitting beside their small pool one afternoon when a boulder came flying at them from a strip job behind their house. That particular piece of flyrock splashed down in the pool, cracked the concrete bottom, and soaked the man and wife. It sounds like a scene from *The Simpsons*, but the shocked couple was slow to see the humor in it. This year, eight off-site flyrock violations were reported in Kentucky, and in one case, children were playing in their pool when debris started falling around them.

In August 2001, Oat Marshall, a blasting supervisor for Consol of Kentucky, resigned rather than set off illegal blasts that he knew would damage property in the small Letcher County community of Deane. He sued Consol and its contractor, the El Dorado Chemical Company, who later settled for $142,500.

But illegal blasting continues all over the coalfields. Clinton Handshoe of Hueysville has had five wells go dry because of blasting. "You have to take the companies to court to prove that they done it," Handshoe told me. "How are you going to get water in the meantime?" And if you don't have the money to hire an attorney, how are you going to take them to court?

Down below, the driver of a blast hole drill is slowly working his way around the perimeter of this bench, boring a sixty-foot-deep hole about every ten feet. With a long, vertical drill carriage attached to the chassis, the machine moves with the slow unsteadiness of a man carrying a long ladder. I can hear a low grinding sound as the hydraulic motor builds torque and the long drill bit tears away at the sandstone.

While the rock drill moves from hole to hole, each one seven inches in diameter, two men step out of the truck that carries the explosives. A long tube that stretches over the top of the tank now swings to the side, where one of the men holds a narrow plastic bag up to the mouthpiece. A brownish substance fills the bag with a concoction known in the coal industry as ANFO. The acronym stands for ammonium nitrate and fuel oil.

Too volatile to transport, the ammonium is mixed with diesel fuel on the mine site. The two men drop the mixture down into one of the blast holes, then repeat the process around the edge of the bench. Finally, they pack blasting caps—detonators—into each hole and string them together with a long orange fuse. Each blasting cap has a timer built into it. As their truck and the rock drill pull away from the bench, the warning siren sounds. Each permitted mining operation must follow something called the scaled-distance equation, which I don't pretend to understand. It reads like this:

SD (ft./lb.) = distance (ft.) / charge weight (lb.) ½

The gist of it is that the distance from the nearest residential or commercial structure dictates how many pounds of explosives can be set off. Regulators make quarterly inspections of the mine sites and use seismographs to measure each blast. But the problem, coalfield residents will tell you, is that all the inspections are announced, which makes it rather easy for a coal company to exceed the blasting limits as soon as the inspectors are gone.

When this particular blast finally goes off, it looks something like a Las Vegas fountain suddenly coming to life. Except it is spewing rock instead of water. All at once, white plumes of debris shoot out of every hole, and then seem to hang for a moment at about thirty feet. Two seconds after the blast, the entire outer ledge falls away, as if it had been shaken by an earthquake, which of course it has. I duck and cover as smaller debris scatters in my direction. An acrid yellow smoke fills the air and a fine gray powder settles over this section of the mine site. Slowly, the haul trucks and front-end loader move in to cart off all this loosened rock. From this perspective, it is easy to understand why Teri Blanton calls mountaintopping "bomb and bury." *Removal* is certainly too clinical, too surgical a term. All this rubble will be dumped down into the valley below, and this same process will be repeated thousands of times across this mine site.

I climb back up to the crest, then drop down the backside of Lost Mountain. Here the ridgeline forms another large bowl, and all sides bear the crisscrossed scars of bulldozers that have leveled most of the trees. Two cardinals call back and forth across the valley. When I move too close, a squawking ruffed grouse abandons its pile of leaves and disappears down into the hollow. A little farther on, I find one source of its distress. A pile of brown and white feathers is all that's left of another grouse, probably the victim of a bobcat or a coyote.

My T-shirt snags on the briars of young black locust trees. I pull myself up over a series of bald escarpments where the sun has cast an almost lavender hue over the sandstone. Then, abruptly, I stumble into a clearing, an artificial shelf carved out by a bulldozer. An uprooted maple has shot out, for the last time, its long scarlet stamens. Other trees lean away, half unearthed. Above me rises a thirty-foot mound of debris. Loosened by blasting, then pushed aside, this rock-and-shale mound forms a rim around the entire mine site. I climb up over the boulders and buried tree limbs until I am crouching at the edge of this cratered landscape. It is, believe me, no flight of the imagination to say it looks from here as if a meteor has hit this side of this mountain. A front-end loader moves across the deepest section, a black pit cut into the number 10 coal seam. The operator scrapes the bituminous rock into piles, waiting for the coal trucks to return for another load. A flat gray plateau stretches out to the hollow fill, where haul trucks dump their endless loads. At the edge of the hollow fill, a narrow road winds up to the next shelf cut into the side of the mountain. Above that, another bench has been leveled about forty feet from the summit. A flat, vertical escarpment, a high wall, rises behind these benches, and at the bottom of each high wall runs a four-foot-thick seam of coal. From this spot, I can see each of the three highest bands receding and ascending up the mountain. Back in the '50s and '60s, most companies would have augered the periphery of those seams and been done with it. But modern explosives, larger equipment, and a rapacious public that uses 70 percent more electricity than it did twenty years ago mean that the entire mountaintop must go.

WHAT IS A FLYING SQUIRREL WORTH?

Naturalist Jim Krupa was nearly sprinting up a mountainside in Robinson Forest, a twelve-thousand-acre woodland that sprawls across three counties. Behind him, a group of his University of Kentucky biology students and I panted and grasped for roots to pull ourselves up the steep incline. "Almost to the top," Krupa called back for the third time.

We were after the Southern flying squirrel, a nocturnal aviator. The day before, our group had descended this same ridge, stopping at ten evenly spaced trees along four different elevations to set a total of forty #102 Tomahawk live traps—wire cages the size of a shoebox. The students wrapped the traps in plastic—a dry squirrel is a happy squirrel—and duct-taped them to wooden ledges that Krupa had nailed six feet from the ground. In the back of each trap was placed a handful of cotton, for bedding, and a meatball-sized mixture of oatmeal and peanut butter. The only thing missing was a Gideon Bible.

When the students and I finally reached the capstone the next morning, Krupa was cleaning his glasses with the end of his T-shirt. He is a short man, barrel-chested and good-humored. He has been trapping flying squirrels in this forest for almost twenty years, and through a lot of trial and error, he has discovered what kind of food they like and what kind of forest they prefer. The day before, Krupa had promised us something spectacular—a squirrel in flight.

As the students and I quaffed water and tried to catch our breath, Krupa started into a kind of stand up routine he has developed over the years to accompany these field trips. "If you set forty traps," he said, "you're an apprentice baiter. If you set a hundred traps, you're a journeyman baiter. If you set over a thousand traps, what are you?"

"A master baiter," answered one of the female students.

Krupa broke into a grin. This was his kind of audience. "I told that joke to another group," he said, "and later a woman wrote me a nasty letter telling me how offensive it was." He seemed genuinely aggrieved.

"Lighten up," someone told the absent accuser.

We started for the first trap. It was empty, its door unsprung. With a pocketknife, Krupa cut the cage from its perch and dumped the cotton balls on the ground. "Why did I do that?" he asked.

There was a pause and then one student said, "So birds can use it for nests?"

"Right."

Krupa dropped the trap into an army-issue backpack one student was shouldering, and we moved on.

Before we reached the second trap, we could see the plastic was tattered. The door was down. At the very back of the trap, an animal the size of a chipmunk huddled on a bed of cotton balls. With his pocket knife, Krupa cut away the duct tape and held up the trap. Much smaller than a gray squirrel, this silky rodent sprang from one corner of the trap to another. Then it froze for a few seconds to assess the situation. We were gazing upon the only gliding mammal in North America. Its brown fur was streaked with black. It had large dark eyes, a pink nose, and attentive ears. A thin black line of fur ran along the edge of a loose fold of skin that extended from the wrist of each forefoot to the ankle of each hind foot. The squirrel seemed to wear it like a cape. The students and I all let out inarticulate sounds of admiration.

"Is it a male or a female?" Krupa asked.

As if on cue, the squirrel jumped to the top of the cage, exposing its furry white underside and two prodigious lumps between its hind legs.

"Holy shit!" exclaimed one of the female students.

"I've *got* to get a picture of that," said another.

In breeding season, Krupa explained, a male's testes can make up 10 percent of his entire body weight. "It has a bone in its penis shaped like an ice cream scoop," he offered, "probably to scrape another male's sperm out of a female's birth canal." As far as Krupa is concerned, everything in the woods, from the bright red skin of a toxic salamander to the penis of a flying squirrel, can be explained by Darwin's theory of evolution. Back at the University of Kentucky, where he teaches huge Intro to Biology classes, Krupa goes out of his way to pick fights with creationists. It has made him one of the more controversial and one of the most popular professors on campus. No one sleeps in his 300-student lecture hall.

"Who wants to do the honors?" Krupa asked this much smaller class as he holds out the trap.

The girl who had shown a priapic interest in this particular squirrel reached for the handle.

"Find a good tree," Krupa told her, "one without any snags."

She decided on a tall red maple and positioned the trap lengthwise against the bark, with the door pointing up.

"Everyone, get ready," Krupa called out.

The student pulled open the door and the squirrel shot up the tree, almost out of sight. At about thirty feet up, it suddenly stopped and spun around. Quickly the anxious squirrel surveyed his options. Then he leapt. That small brown spot near the top of a tree instantly became a flat white glider, soaring above our heads. With the extra web of flesh flung open between his extended legs, *Glaucomys volans* rode the air of the understory as if flying were the most natural thing a mammal could do. He soared about twenty feet, then quickly lowered his flat, rudder-like tail as he caught onto the bark of a larger tree. He had lost about fifteen feet of elevation, and this time he scampered up the backside of the tree, up into the canopy.

"Keep watching," Krupa advised, "he'll probably fly again."

About twenty seconds passed, and then suddenly the improbable aviator shot from the leaves. With his left wing tilted skyward, he swept around two younger trees in a magnificent, careering glide. Then, leveling out, he soared down the ridge side, finally landing right below a snag on a black gum tree, into which he disappeared.

Amazed, the students and I uttered the usual banalities of amazement.

Krupa grinned, proud of his squirrel. "Now," he said, "you've seen something ninety-nine-point-nine-nine-nine percent of the population will never see."

One obvious reason is that *G. volans* is a nocturnal flyer. Another is that most of the human population doesn't know how to trap a flying squirrel, and probably doesn't care to know. But there is also another, far more pressing reason. Human populations are rising and forest populations are falling.

There is an obvious and direct correlation between the overpopulation of *Homo sapiens* and the declining population of just about every other species that has not fallen under human domestication. Around 8,000 years ago, near the dawn of agriculture and after the northern retreat of the glaciers, the world's forests were at their fullest. But with the onset of agriculture, urbanism, and the timber industry, forests have been disappearing ever since. Nearly all of central Appalachia had been clear-cut by the 1920s. And by the time the mountains had regenerated into a diverse second-growth forest, the technology and the economic power structures had been set in place to unleash strip mining across these hills. Today, 95 percent of North America's mixed mesophytic forests have been degraded by species loss. The Cumberland Plateau is one particular "hot spot" of ecological concern, where over two-thirds of the songbird population is on the decline.

Throughout these chapters, I've mentioned several eastern species whose numbers are falling quickly because of forest fragmentation, much of it caused by strip mining. But the problem cannot be overemphasized, and the Southern flying squirrel shows why. In the ten traps set along that highest transept, we caught three more squirrels—one more male and two females, one pregnant. Over the years, this has proven to be Krupa's most successful trapline. The trees grow close together and in great variety. *Glaucomys volans* has lots of enemies—raccoons, rat snakes, barred owls—and as a result, it is sensitive to habitat. The Southern flyer wants a mature forest with lots of older trees into which she might escape predators. She wants plentiful hard mast—hickory nuts, beechnuts, acorns—near her nest, usually a tree cavity hollowed out by a woodpecker. She does not want to range far for food, and she particularly does not want to linger near a forest edge, where predators abound and she is easier to spot.

For all these reasons, this particular ridge side, well ensconced within Robinson Forest's contiguous twelve thousand acres, plays host to a large number of flying squirrels. Once, with his forty traps, Krupa trapped thirty-eight squirrels in a single day. By contrast, in a study published last January in *Mammalian Biology*, J. F. Taulman and K. G. Smith found that male flying squirrels who had to forage farthest from their home range were also the ones that lived in fragmented forests. Of seventeen squirrels collared in such an open forest, only two survived the following summer.

The difference between that study and Krupa's more informal findings reflects a global trend in species decline due to the evisceration of forest communities. In his remarkable book *Song of the Dodo*, David Quammen has traced the massive rate of extinction caused by forest fragmentation all the way back to Alfred Russel Wallace and the discovery of the theory of evolution. Like Darwin, with whom Wallace shares the discovery, the younger Wallace reached his own conclusions through the study of islands—in his case, the Malay Archipelago north of Australia. What he found was that two tigers who once roamed, say, the peninsula of Malaysia, could become completely different species if rising sea levels separated them by distance, and a million years separated them by time. Isolation plus time created species divergence. In other words, the key to evolution lay in the study of islands, from which we get the key term *island biogeography*. And the larger the island, say Madagascar, the larger the diversity on that island. Size matters.

Now flip the theory around, as did E. O. Wilson and R. H. MacArthur. If older and larger islands give rise to a greater diversity of species, then creating an island effect among forests—that is, turning one large forest into many smaller woodlots—does just the opposite. According to Wilson and MacArthur's species-area curve, the greater the area of a forest island, the greater the number of species that will exist within that forest. And the smaller the area, the greater the chance of smaller populations among species, more inbreeding among species, more predation from the edges, and more parasitism by species like the cowbird, who would rather someone else raise her young. The upshot: We are currently witnessing—and ignoring—the sixth great extinction since the advent of life on earth. This is not the hysterical cry of some druid; it is cold scientific fact.

The history of life on earth has seen five great extinctions, all caused by natural phenomena. The last extinction, 65 million years ago, possibly caused by an asteroid, wiped out the dinosaurs. Everybody learns this in school. What we don't usually learn is that up until the age of agriculture, species became extinct at roughly the rate of a few every million years—the same rate at which new species evolved. Or as E. O. Wilson has figured it, one species out of a million went extinct each year. That formula is known to biologists and archaeologists as the "background rate." Now, because of rising temperatures, chemical pollution, the introduction of exotic plants, and forest fragmentation, species worldwide are disappearing at 1,000 to 10,000 times that rate. That is to say, roughly one species goes extinct every hour.

According to the World Conservation Union Red List, one in four mammals and one in eight bird species are in some degree of danger. Since 1980, habitat destruction has reduced our closest genetic and socially predisposed relative, the bonobo of West Africa, from 100,000 to merely 3,000. Of the 9,946 known bird species, 70 percent are declining in number. Wilson concludes, "If the decision were taken today to freeze all conservation efforts at their current level while allowing the same rates of deforestation and other forms of environmental destruction to continue, it is safe to say that at least a fifth of the species of plants and animals would be gone or committed to early extinction by 2030, and half by the end of the century."

Even Robinson Forest and its flying squirrels are not safe from the bulldozer's blade. There are millions of dollars' worth of coal beneath these trees and capstones. The University of Kentucky owns the forest, and in an age of lean state budgets and painful cuts in higher education, harvesting some of that coal can seem to some like an obvious panacea. The university's current president, Lee Todd, has told the board of trustees that he has no plans to mine Robinson Forest "at this time," and that last phrase keeps researchers like Krupa anxious about the forest's future. As we walked the highest trapline, Krupa told me there had been discussion among the board of trustees about leveling that particular ridge where the flying squirrels were thriving.

There is a stock metaphor among conservationists when the talk turns to logging or strip mining. It is the analogy of a forest as library: The rain forests are like the great library of Alexandria. Burn off a forest and you might as well have burned the last surviving copies of

Aristotle and Maimonides. Knowledge is being lost, irretrievably in many cases. Consider that one in every ten plant species contains anti-cancer compounds. In a purely selfish sense, *Homo sapiens* who care about the survival of their species should find the current rate of extinction rather alarming. I may think—I do think—that preserving species diversity enriches the very concept of life, but it also holds the secrets to the perpetuation of human life. Because in the end, the natural world does not need conserving. The planet has survived five extinctions; it can survive another one. No, it is *we* who need conserving. And it is the current logic of short-term profit, affluence, and convenience that has blinded us to the havoc we are inflicting on future generations.

We caught ten flying squirrels on that July day in Robinson Forest. Their flights were all different and remarkable. As we stuffed the final cages in backpacks and walked back to Krupa's truck, I asked for his prognosis of the human condition.

"Oh, I think we're doomed," he offered cheerfully. "With our levels of population and rates of consumption, it's just a matter of time before we kill ourselves off." He paused, wiped his glasses on his T-shirt, and smiled. "It's not something I tell my freshmen."

April 2004

LOST MOUNTAIN

Franz Kafka's short story "Before the Law" begins like this: "Before the Law stands a doorkeeper. To this doorkeeper there comes a man from the country and prays for admittance to the Law. But the doorkeeper says that he cannot grant admittance at the moment." And behind this doorkeeper are many others, all guarding the Law. "These are difficulties the man from the country has not expected," writes Kafka; "the Law, he thinks, should surely be accessible at all times and to everyone." The man waits for days, which turn into years. At times he tries to bribe the doorkeeper, who accepts the money with this fateful remark: "I am only taking it to keep you from thinking there is something you haven't tried." In the end, the man dies, and with finality, the doorkeeper shuts the door.

Now, "Kafkaesque" is not a term I throw around lightly. But if you go to enough public hearings on surface-mining legislation, it soon becomes clear that what you are watching is a drama of grand futility straight out of Franz Kafka. And the actors know it. The officials from the Office of Surface Mining sit stoically, almost indifferently, at a table in the center of the stage. One after another, coalfield citizens step to the podium to have their say about the effects of weakening regulations on strip mining, and one after another, they announce to anyone naïve enough to believe in participatory democracy that this is all a done deal anyway. Yet still they put themselves through this compulsory charade. The stenographer who sits at the side of the stage, taking it all down, is himself a character Kafka would have particularly loved—the man endlessly writing a report that no one will ever read.

Kafka, it is sometimes said, wrote the totalitarian history of the twentieth century before it happened. It remains to be seen if he did not also write the history of the United States in the twenty-first century. After all, one of the spookier things about the current state of American democracy is that citizens have the freedom to speak, but unless they are wealthy, not the authority to be heard.

Consider, for instance, the case of Steven Griles. It has been well documented that Griles worked as a lobbyist for the coal and oil industry before he was tapped to be George W. Bush's deputy secretary of the interior. During each year of his term at Interior, Griles has received a $284,000 deferred-compensation package from his former employer, National Environmental Strategies (NES). The *Washington Post* reported that Griles met three times with the National Mining Association (NMA) , a former client of NES, while NMA was seeking looser standards on mountaintop removal. Which is exactly what NMA got. Since the 1977 Surface Mining Control and Reclamation Act (SMCRA) states in rather plain language that mining permits can be granted only if "no damage will be done to the natural water-courses," the Department of the Interior under Griles proposed a rule to "*clarify* [my italics] the circumstances in which mining activities . . . may be allowed within 100 feet of a perennial or intermittent stream." The "clarification" requires that "the mining operation has been designed, to the extent possible, to minimize impacts on hydrology, fish and wildlife . . . prior to allowing mining within 100 feet of a perennial or intermittent stream." It's the phrase *to the extent possible* and the word *prior* that could render the stream protection of SMCRA unenforceable.

"Steven Griles is a monster," former mining engineer Jack Spadaro once told me flatly. "Under Reagan and the first Bush, it was he more than anyone who set in motion all of this mountaintop removal."

In a more sensible time and place, a citizen might logically ask why someone who has spent his adult life protecting and profiting from the coal industry would then be allowed to regulate it. Beyond that, the citizen might ask why a government agency allegedly designed to protect the people does not protect one of their most vital needs—water. And that citizen might receive an answer.

I don't mean to suggest that the Democrats were much better at enforcing strip mining laws. From the time Jimmy Carter signed SMCRA in 1977 to the time of Ronald Reagan's election three years later, there was a legitimate zeal on the part of regulators to hold coal operators accountable. Office of Surface Mining field manager Patrick Angel told me, with obvious pride, that in 1977 he issued the first cessation notification at the Cloverfork mine in Harlan County. But by the time Bill Clinton took office in 1993, that crusader attitude had mostly vanished. As Martin County resident Mickey McCoy likes to say, "The watchdogs have become the guard dogs of the industry." And Clinton did nothing to change that. According to longtime environmental lawyer Tom FitzGerald, he largely abandoned his environmental agenda after Newt Gingrich's "Republican revolution" of 1994. As a result, the coal industry has spent the last thirty years finding ways to weaken or ignore SMCRA.

Given Griles's new proposal to lawfully bury streams, I decided it might be a good time to look for the headwaters of Lost Creek before they too disappear or die.

I pull off the road beside a small house that sits closest to the mining on Lost Mountain. Chickens strut around a neat backyard, and behind the chickens, I can hear the low voice of Lost Creek. I follow it up a narrow gorge, stepping over cobble and fallen tree limbs. Two days of rain have brought the stream to life. Chickweed and rue anemone bloom modestly along the creek side. Bent trillium is about to spread its maroon petals. Redbud and dogwood flower all through the understory.

There is something I have long admired about the attention ancient Chinese poets paid to particular elements of a mountain landscape, and the way that just naming those particulars was enough. A flower didn't have to stand for some abstract human emotion; it could simply be itself. Here is a characteristic poem by Wang Wei:

Men sleep. The cassia blossoms fall.
The Spring night is still in the empty mountains.
When the full moon rises,
It troubles the wild birds.

From time to time you can hear them
Above the sound of the flooding waterfalls.

It is as if the poet and the natural world have met halfway in the poem's images. And for the reader, those images become thresholds back into a natural world we are often too distracted to see.

So it was a minor epiphany for me to learn that the forests of Appalachia are almost identical to those in southern China where Wang Wei, Li Po, and Tu Fu wrote their famous poems about cascading waterfalls and full moons. Two-thirds of all the wild orchids in Appalachia are cousins to those in China. There are only two species of tulip poplar in the world—one in China and one in the eastern United States. In both forests, the nonwoody plants have developed underground storage systems, and most of them bloom this time of year, before the canopy closes above them. Inspect the forest floor in southern China and Appalachia and you will find the same mayapple and jack-in-the-pulpit, the same ginseng and ferns. Apparently, what connects these two ecosystems on opposite sides of the globe is that neither suffered extensive glaciation during the Pleistocene era. When the ice withdrew, only these two regions retained the plant diversity that was once characteristic of each entire continent.

Halfway up the gorge, on the backside of Lost Mountain, the trickling stream disappears beneath a large beech tree surrounded by rhododendron. I take a seat on a fallen branch to contemplate Lost Creek's modest beginning. There is something intrinsically rewarding about finding the source of any river. Here is one particular creation story, and you can bear witness to it. And this is the story in its purest form. There have been no redactions, no spurious insertions. This is a story you can trust. It comes straight from the source.

What is not as rewarding is to hear machinery rumbling above the source of that stream. I climb the creekbed until I can see a backhoe prying loose boulders and subsoil up at the head of the hollow. It is working in tandem with a dozer, slowly extending a bench along the backside of the mountain. The backhoe chips away at the rockface with its long mechanical arm, and the dozer pushes the debris aside. According to the mining maps for this job, none of the spoil is supposed to be deposited in this creekbed. But last week, under the Freedom of Information Act, I elicited fourteen single-spaced pages of violations by Leslie Resources. Since 1985, Leslie has racked up over five hundred citations. Forty-seven of those violations pertain to water quality, while twenty-four are for illegal blasting. Leslie Resources is particularly lax about keeping sediment out of streams. Here on Lost Mountain, the company has already been cited for allowing water to wash off the mountain without proper sediment control.

The problem extends all across Kentucky. A recent study conducted by the state's Division of Water found that 47 percent of Kentucky's rivers and streams are too polluted for drinking, fishing, or swimming—a figure that has risen 12 percent in the last four years. The study also found that the largest source of the pollution is sedimentation, and most of that run-off is caused by mining. And in the coalfields, matters are much worse. Biologist Greg J. Pond of the Kentucky Department of Environmental Protection found that 95 percent of the state's headwater streams have been impaired by surface mining. Sulfates are produced during the strip-mining process, and those heavy metals—calcium, magnesium, and iron—wash down from the hollow fills and into streams. These dissolved solids decimate nearly all species of macroinvertebrates, such as mayflies, a key member of headwater-stream communities. Pond points out that these headwater streams serve as "capillaries" that provide high levels of water quality and nutrients to downstream communities, including human communities. Pond also found that sedimentation pollution (i.e., siltation) is the number-one stressor to aquatic life in Kentucky, and surface mining is the leading cause of sedimentation. In addition, that sedimentation is pushing streams to over half of their storage capacity, and that will inevitably cause more flooding and greater cleanup costs downstream. Consider the example of New York City. In 1989, the EPA ordered the city to build a water-filtration plant that would cost $8 billion to build and $300 million a year to maintain. Instead, New York spent $2 billion reforesting a 2,000-square-mile watershed in the Catskill Mountains. For billions of dollars less, that riparian forest resumed the purification tasks it had been performing for millions of years. The same would be true in Kentucky, where sedimentation is particularly expensive to filter out of drinking water. Yet I would wager that most Kentuckians living in Louisville have no idea of the connection between the costs they will pay for clean tap water and the loss of eastern Kentucky forests, nor do most of them understand the relationship between that cost and their increasing use of coal-fired power.

But farther upstream, problems are more serious than water bills. An Eastern Kentucky University study found that children in Letcher County suffer from an alarmingly high rate of nausea, diarrhea, vomiting, and shortness of breath that can all be traced back to sedimentation and dissolved minerals in their drinking water. Long-term effects can include bone damage, cancers of the digestive track, and liver, kidney, and spleen failure. Erica Urise, who lives on Island Creek in Grapevine, Kentucky, told me she has to bathe her two-year-old daughter in contaminated water. "She loves baths," Urise said, "but I can't let her play in the tub with any toys that she might drink from. I have to give her a sippy cup full of juice so she won't drink the water."

To avoid the dozer and backhoe, I cut a wide tack around the backside of the mountain. Near the ridgetop, ground pine (*Lycopodium obscurum*) has begun to poke its bright green fronds up through matted leaves. This coniferous-looking fern tops out at seven inches, but its distant ancestor *Lepidodendron* grew to 150 feet as it breathed in vast amounts of carbon. That was over 300 million years ago. Oddly, though, these trees that looked like giant ferns often fell into oxygen-poor bogs, and so they never decayed. Instead molten heat and geological pressure hardened them into the compressed layers of black, carbon-rich rock that is disappearing fast from the other side of this ridge.

When I reach the mountaintop, I discover how fast. The dirt road that led along the eastern ridge of Lost Mountain is gone, and so is the eastern ridge. What was once an arching razorback is now a sunken crater. An explosion is set off inside the deep pit, but I see only the tops of the gray blast plumes. The source of Lost Creek lies right below this pit, just over the ridge. What this blasting will do to the groundwater might not be fully understood for several years. What is known is that when underground pirite is oxidized through blasting, it releases sulfuric acid. And it is almost certain that blasting on Lost Mountain will create underground fissures through which mine acid will drain down into seeps that will leach out into this watershed. Because of acid mine drainage, along with acid rain, the only trout native to Appalachia, the brook trout, is on the decline. And the iridescent brookie is fairly resistant to acidic waters. If its populations are sinking, so are those of nearly all the aquatic life in mountain streams.

Perhaps the problem that mining causes along these waterways can best be measured by driving about fifteen miles to the Falling Rock Watershed in Robinson Forest. That watershed feeds one of the cleanest streams in Kentucky, Clemons Fork. Its level of conductivity—that is, dissolved solids—is usually between 50 and 60. Its chlorides, sulfates, magnesium, and sodium levels all hover around one milligram per liter of water. But one has only to go a half-mile downstream, to the confluence of Buckhorn Creek, which sits directly below a strip mine, to find the conductivity has risen to 1,000, the magnesium and calcium to 25, and the sulfates from less than 10 to 300. Whereas Clemons Fork can sustain roughly a hundred species, water conditions at Buckhorn Creek have been so severely degraded that at most ten species can survive.

ACTS OF GOD

In 1912, the railroad finally reached McRoberts, a small hamlet that sits near the headwaters of the Kentucky River's north fork. The Northern coal barons had been waiting for decades to get at the minerals in the Cumberland Highlands, and once the tracks were laid, they quickly threw up coal camps along the narrow valley floors. Elkhorn Coal built nearly a hundred four-over-four houses about ten feet from either side of the only road that leads to McRoberts. Two families lived in each house. There was no indoor plumbing.

Men who had tended their marginal farms traded plows for picks and went to work in the new mines. They made enough to pay rent to the coal company and buy canned food at the company store. The latter would become a symbol of the miner's loss of independence. "Though he might revert on occasion to his ancestral agriculture," wrote Harry Caudill, "he would never again free himself from dependence upon his new overlords." It was in a very real sense that the miner narrating the song "Sixteen Tons" felt that he owed his soul to the company store.

Today, many of the company houses are still standing in McRoberts, covered with tar shingles or vinyl siding. Interspersed between them are buildings of brick and cement blocks that house numerous congregations of the Free Will Baptist Church, the Regular Baptist Church, the Old Regular Baptist Church, and the Primitive Baptist Church. Driving up the hollow last summer, I calculated a ratio of about one church for every twenty homes. For members of a denomination that believes in the Bible's inerrancy, these congregations still seem to have found a lot to disagree on. Some sociologists contend that the number of churches in Appalachia, many Calvinist in nature, reflects an important characteristic of the region—fatalism. In this view, such seeming injustices as exploitive bosses, terrible working conditions, and grinding poverty may not get righted in this world, but will certainly be compensated for in the next.

In a sense, it was that quietism, particularly on the part of the church, that brought me to McRoberts. Though I was raised Baptist, I had not actually been in a Baptist church for almost two decades. Martin Luther King, Jr., said that one of his most bitter disappointments was with the mainstream church, which gave lip service to the civil rights movement but took little action to bring about change. That pretty much became my position when I moved out of my parents' house. But I had come to McRoberts looking for the Reverend Steve Peake, a local preacher who actually had done something about the mountaintop mining that surrounds this valley community.

In 1998, Tampa Energy Company (TECO) started blasting along

the ridgetops above McRoberts. Homes shook and foundations cracked. Then TECO sheered off all the vegetation at the head of Chopping Block Hollow and replaced it with the compacted rubble of a valley fill. In a region prone to flash floods, nothing was left to hold back the rain. As a result, this once-forested watershed turned into an enormous funnel. Between the blasting and the flooding, the people of McRoberts have almost literally been flushed out of the hollow. Larry Easterling and his mother were flooded out five times in three years. Three so-called hundred-year floods happened in ten days. The residents of Choppin Branch Road, which sat right below the valley fill, had a foot of mud in their homes. After that, Letcher County Judge Executive Carroll Smith drove up to the mine site, where he was met by TECO executives and a state inspector. "I was pretty mad and I said some pretty strong things," Smith told me. The state inspector told the judge that he was "way out of line," that TECO had done nothing wrong. "I told him he should learn the difference between legal and illegal, and right and wrong," Smith recalled. "What TECO was doing might have been legal, but it wasn't right. If I have to make a living by walking on my neighbor, then I'm not going to do it. And if it's legal, then the law needs to be changed."

That part about treating your neighbor right certainly has some biblical foundations, and it was one reason that in December 2002 Steve Peake led a group of fifty people up to what was once the top of the mountain above Chopping Block Hollow, where he said a prayer, asking that the hearts of coal operators might soften. Someone else scattered wildflower seeds over the gray waste. Hymns were sung. A small group of reporters came to write it all up.

In response, the coal industry, which had initially dismissed the terrible flooding of McRoberts as an act of God, now decided, in Bill Caylor's words, that "it's improper to drag the Bible into this debate." But since the Prayer on the Mountain, as it has become known, the Bible has been part of the debate—perhaps the strongest part—and Steve Peake gives regular tours to groups who want to understand the destruction.

The Corinth Baptist Church is a modest white building. I found the Reverend Peake trimming the grass around the church's block foundation, which was visibly sagging. He turned off the Weed Eater and we shook hands. "This is an historic building," he said. "In each coal town, the company would build a church. Elkhorn Coal built this one for the black community. It would be used for school during the week, and then on Sunday we had church here."

Peake's grandfather and father were among the many Southern blacks who, once the coal camps had been erected, migrated to eastern Kentucky to work in the mines. They rented half of a two-story house from the coal company. It had a living room, a kitchen, and two bedrooms upstairs. Peake and his two brothers slept in one bed, and his three sisters slept in another. "We were living in duplexes and didn't even know it," Peake said, smiling. "We thought we were living in half a house."

I asked about the poor condition of the church's foundation. "It's from all the jarring and water damage," Peake said. He showed me into the small, brightly lit sanctuary. "See how the walls are bowing out?" They were indeed leaning from where the foundation had sunk, and the ceiling looked none too stable. "Now, if I show that to the coal company," Peake said, "they'll just say, 'Well, that's an old building. Nothing lasts forever.' And they have lawyers just waiting for your case. A man would have to be retired with nothing to do to complete all the paperwork. A workingman couldn't do it." (It was a complaint I had heard before. At one of the public hearings in Hazard, a man named Dan Cash stood up and complained, "They just run you from one agency to another until you finally give up. That's tyranny, the real thing." And in addition to bureaucratic runaround, the baroque language of the legal drafts and the permit provisions seems designed to intimidate and alienate anyone without a law degree.)

Peake's friend, Lucius Thompson, a retired miner, lives in a trailer near the TECO mine site. The blasting made his foundation sink at both ends, so that the trailer now looks like a half-opened jackknife. "You can't set a glass of water on a table in his house," Peake said. "It will just slide right off." TECO's assessment? You've got an old trailer. Nothing lasts forever.

Peake's congregation first started meeting at the Pleasant Run Baptist Church in McRoberts. But when the blasting started four years ago, a huge tear appeared in the church's roof. "We repaired it numerous times," Peake said. What did TECO tell Peake? You've got an old roof.

So the congregation moved down the road to the Corinth Baptist Church, where the minister had just passed away. Peake, a thin, muscular man, balding slightly, showed me into his sparse office. A quilt depicting the Last Supper hung behind his desk.

The story of Peake's community is the same one you hear all across the coalfields. An outside corporation comes in, hires few if any local people, extracts the minerals, then leaves the community with acid streams, flooding, cracked foundations, and bald hillsides.

In 2001, things were at their worst. The floods came and swept through people's homes. Foundations started cracking and sinking. The water washed away Peake's carport. Then he noticed that the Department of Transportation signs stating a thirty-ton carrying capacity of each bridge had disappeared. The coal trucks were hauling four times that weight over the same bridges that school buses used. But worst of all, said Peake, was the dust. "Every minute and a half, an eighteen-wheeler would go by hauling coal. You couldn't walk down the road without being covered over in dust. You could dust your house in the morning, and in the evening you had to dust it again." His neighbors began to develop serious cases of bronchitis and asthma. Older residents suffered the most. Clinton Handshoe, a man in his late sixties, can write his name in the thick dust that covers his patio furniture. "I can't use the lawn, I can't sit on my porch," he told me. "I'm a prisoner in my own home." When his grandchildren come in from playing, they are covered in coal dust. Then Handshoe's daughter has to bathe them in water contaminated by blasting, if there is any water at all.

One of the saddest stories, alluded to earlier, comes from the head of Chopping Block Hollow, where Debra and Granville Burke lived. First the blasting above their house wrecked their foundation. Then the floods came. Four times, they wiped out the Burkes' garden. "We depended on canning and food from the garden to get through the winter," Granville Burke said. On Christmas morning of that year, Debra Burke took her life. In a letter published in a local paper, her husband wrote, "She left eight letters describing how she loved us all but that our burdens were just getting too much to bear. She had begged for TECO to at least replace our garden, but they just turned their back on her. I look back now and think of all the things I wish I had done differently so that she might still be with us, but mostly I wish that TECO had never started mining above our home."

Around that time Peake and several other clergymen began planning the Prayer on the Mountain. They weren't asking for divine intervention, just a little awareness. "If I'm aware that what I'm doing over here is hurting you over there, then maybe I'll quit doing it," Peake reasoned. "If people can be made aware, I think a difference can be made. Our Prayer on the Mountain was a prayer that these coal operators will recognize what they're doing."

It's an ethic at least as old as the Bible, from which Peake draws the fundamental lesson about the suffering caused to his community: The love of money is indeed the root of all evil. "What's happened here is that greed has caused people to do all kinds of things," he said. "You made a million dollars, but how many families did you destroy? Deep mining didn't disturb the land that much. But when you pull an entire mountaintop back, you're destroying the land. You're destroying the cycle of nature. And I think God is going to hold somebody accountable."

Peake doesn't buy into the argument that since the God of Genesis gave Adam and Eve "dominion" over the earth, humans have a right to use its resources in whatever ways we desire. Peake understands "dominion" to mean "stewardship." "God put us here to take care of the earth," he said. "It's a two-way street. We take care of it, it takes care of us. Look at the Indians. They lived off the land, but they didn't destroy the land. When I was a boy, we would go up in these mountains and they were beautiful, untouched. Now kids go bike riding up a trail, and what used to be more mountain turns out to be a big pit."

For now, things have settled down a little in McRoberts. After the

Prayer on the Mountain and the publicity it received, TECO started hauling its coal down a different route. But Peake knows that more mining is coming. Company helicopters have been flying over McRoberts, scouting future mine sites. And if the weight-capacity signs disappear from the bridges again, Peake plans to do something about it. "Next time I'm going to make some noise," he said. "Sometimes what happens is we think we're too little to make a difference. We think one man's griping won't do anything. I think it's going to take one at a time to get things done."

Who knows? Maybe in the next life, there is a reward for earthly suffering. But what would it hurt to alleviate some of it while we're here?

May 2004

LOST MOUNTAIN

The photos of American torture at Abu Ghraib surfaced this week. I have often felt despondent about decisions that American presidents have made in my name, but this is the first time I have felt truly embarrassed to be an American. I am looking forward to seeing Lost Creek; I am remembering my favorite line from Thoreau: "He who hears the rippling of rivers in these degenerate days will not utterly despair." And when I pull off the main road north of Hazard, the creek is running clear and strong beneath the mixed mesophytic forest, now in full leaf. Families have set out their creekside gardens. Neat rows of early greens are almost a foot high.

This is the ninth month of my covert sojourn on Lost Mountain. I park below the eastern slope and start up toward the headwaters of Lost Creek. I have taken to wearing camouflage pants and earth-tone

T-shirts on my visits, and so far I have avoided the attention of Leslie Resources.

Canopy leaves have now closed over this gorge, turning the air cool and moist. They have also muffled the sound of the large machines over the next ridge. As I step deliberately over the slick stones, an unannounced explosion makes the entire ridge side tremble. But it is the mental shock more than the physical tremor that knocks me off balance, causing me to fall against a patch of ferns. I right myself and start climbing up the left bank, toward a clearing that affords a profile of the mining. From that standpoint, at about a thousand feet, the mountain looks like a hideous wedding cake, a series of black and gray ledges that lead up to the summit, now only a rocky knob. There, an abandoned cinder-block shack still stands like some ominous cake decoration, covered in graffiti that bears the promising sentiment MIKE LOVES ME BITCH.

From this vantage, an invisible trajectory runs up the eastern slope and divides Lost Mountain into two stark economies—that of the strip mine and that of the broadleaf forest. And from this perspective, what is valued by each economy is easy to discern.

Two empty explosive boxes have been discarded on the bench level where I'm standing. I'm moving in for a closer look when the sirens go off on the other side of the highwall. I quickly retreat to my earlier position, reasoning that if this is a legal blast, I should be all right. Fortunately, it is. A black spray of rock comes shooting over the highwall, but the closest chunks land about thirty feet away. Still, higher ground seems to be what's needed.

I drop down into the watershed, where all the leaves are covered with the chalky gray residue of blasting, then I follow my usual climb up the backside of Lost Mountain. Near the peak, chestnut oaks dominate the canopy. Sassafras and redbud fill in the understory, where a cool breeze is moving. I step around foamflower and bright red catchfly, so called because its sticky stem slows down insects to guarantee a fair exchange of nectar for pollen.

Though this side of the forest is quiet, I notice a silent ovenbird eyeing me from a low twig about thirty feet away. He has a handsome brown head, similar to that of a wood thrush, but his white breast is streaked with black instead of spotted like the thrush's. This neotropical migrant has probably just returned to its breeding ground. The males reach the Eastern forests about two weeks before the females to establish territory. He is usually an ardent suitor, his habits made famous by Robert Frost's poem "The Ovenbird":

There is a singer everyone has heard
Loud, a mid-summer and mid-wood bird . . .

The male takes up his all-day *teach-er teach-er* song once the female's eggs have incubated. For her part, the female ovenbird builds a brilliantly camouflaged watertight nest on the ground. But because of this precarious placement, she likes at least forty acres of continuous forest cover to improve her brood's chances against snakes and owls.

Biologists speak of "indicator species," those that can tell us something important about an ecosystem. In Frost's poem, the ovenbird is indicative of lateness—lateness of season and lateness of the human industrial age.

He says the early petal-fall is past
When pear and cherry bloom went down in shadows
On sunny days a moment overcast;
And comes that other fall we name the fall.
He says the highway dust is over all.

Frost slyly suggests that "that other fall" is both the natural season of dying and the human separation from a prelapsarian state of nature. And then the machine suddenly enters the garden, kicking up dust—in this case, the dust from coal trucks and ANFO blasts. Finally, Frost's ovenbird becomes an indicator in a final sense:

The bird would cease and be as other birds
But that he knows in singing not to sing.
The question that he frames in all but words
Is what to make of a diminished thing.

The ovenbird's song is a eulogy, not a jubilate. In singing, he knows there is less and less worth singing about. And so he poses the crucial question: What is to be made of a diminished thing? The answer, of course, lies just over this ridge.

From the summit, I ease down the southern slope around boulders and a stand of wild azaleas covered with nodding orange blooms. I take up a position behind the largest chestnut oak still standing on this side of the mountain. Two feet beyond it, a highwall drops about seventy feet straight down to the number 11 coal seam, which is now a flat black plateau, stretching out like tarmac. Back at the EIS hearing, one man had stepped to the microphone and asked, "What are these mountains good for? They're all up and down." He would be pleased with what has transpired here on Lost Mountain, where a pilot could easily land a small prop plane on the wide, level shelf below. As it is, two front-end loaders are filling the bucket of a coal truck from both sides. When they are finished, a long mechanical arm pulls a red, white, and blue tarp up over the coal. The truck pulls away and another takes its place. Since I started coming to Lost Mountain, the price of coal per ton has jumped from $34 to $55—coal prices always follow oil—and the pace of its extraction has quickened.

One of the permit maps drawn up for this particular job shows the "pre-mining" contour of the mountain as a dotted line—something almost hypothetical, arbitrary. The "post-mining" contour is designated by two dark lines, flat as a dead man's EKG. When I first looked at that map, it seemed so impossible. Over two hundred feet lay between the dotted outline of the mountaintop and the flat line that indicated a reclaimed "pasture." Didn't the engineers know this was solid rock up here? Didn't they know this ridgeline had been standing longer than the Himalayas? Now, of course, I see they knew that perfectly well, and they knew exactly what they were doing. I had made the mistake of thinking in geological time. But as Rachel Carson wrote in *Silent Spring*, "In the modern world, there is no time." It has been annihilated by explosives and fossil fuel and hydraulic rock drills.

The pit that had been blasted out of the eastern ridge last month is now a gigantic black gash that looks like the work of an earthquake and opens like a canyon onto the southern side of the mountain. Around on the western side, all that's left is a pocked, deracinated landscape, strewn with boulders and absent of anything that could be mistaken for life. Off in the distance, I count nine pickup trucks.

When the 4:30 whistle has blown and those pickups have disappeared down the mountain, I circle around to the nearest bench. From here, the highwall reaches forty feet up to the summit, where a clutch of pine trees hangs over the precipice. I climb over the rubble down on the western side, then follow the lower, longer highwall that sits above the number 10 seam. Because the landscape shifts so quickly beneath the force of the explosives and dozers, a sense of vertigo sets in as I wander around these unnatural formations. Where last month I walked a ridgeline, this month, in exactly the same place, I'm standing on a black plateau, and it's hard to even remember what the original contour looked like. I know it was here, I know there were a few trees left. Now there's nothing. Everything that once stood here now lies a hundred feet away, down in the massive hollow fill. I stretch out my arms and slowly turn full circle. My throat tightens and my breath suddenly becomes short. I cannot see one living thing.

WHITEWASH IN MARTIN COUNTY

Inez, Kentucky, used to be Eden. Literally. The rugged men and women who founded this hamlet near the West Virginia border felt inspired enough to name it after the biblical garden. But as it turned out, there was already an Eden in the western part of Kentucky, and a lot of mail was getting routed in the wrong direction. So the postmaster of the eastern Eden renamed his post office box after his daughter Inez, and the name stuck.

What is wonderful about this story is how it reveals what people once thought about the central Appalachian mountains. It wasn't a place of sheared-off mountaintops, dirty streams, and trash-lined roads; it was a landscape worthy of comparison to an earthly paradise. Today, as you drive into Inez, you see looming overhead the guard towers of a prison built on a former strip mine and steep, unreclaimed valley fills waiting to wash down across the highway. One family's home sits directly beneath a gray slope of rubble that rises a hundred feet behind it. A flash flood in the middle of the night could easily cause a landslide that would crush everyone in the house. I stopped at the mine entrance to write down the name of the mining company and its permit number—Martin County Coal, 800-0149—but really to no purpose. This kind of mining may be criminal, but it isn't illegal.

What Inez is known for today, if it is known at all, is the nation's largest man-made environmental disaster east of the Mississippi. In October 2000, a coal slurry impoundment broke through an underground mine shaft and spilled over 300 million gallons of black, toxic sludge into the headwaters of Coldwater Creek and Wolf Creek. When coal is cleaned for market, the remaining waste and cleaning compounds are stored in huge impoundment ponds, where the heavy debris settles on the bottom, forming a thick sludge. There are 650 such ponds in Appalachia; 225 of those are in Kentucky. On October 11, 2000, the bottom of an impoundment pond that sat above Inez on a Martin County Coal job gave way, and gelatinous black waves of slurry spread over the valley community of Coldwater Creek, on their way to the Ohio River. The sludge moved with the speed and the consistency of volcanic lava, choking everything in its path. Yards and gardens were buried. Bridges were swept away. Basketball goals looked like buoys in a hellish ocean. The only thing people had to be thankful for was their lives. The 1972 Buffalo Creek pond break in West Virginia killed 125 people, but no one died that day in Martin County. The 2000 spill was almost three times the size of the Buffalo Creek break, however, and it was thirty times the size of the *Exxon Valdez* disaster, though you wouldn't have known it by reading the *New York Times*. For months following the Martin County disaster, the

Times didn't print one word about it. One Martin County resident finally concluded, "We're just not quite as cute as those otters." In other words, Prince William Sound was a pristine waterway. But the Appalachian Mountains and their people were already considered damaged goods.

Still, if one pond break was responsible for 125 deaths at Buffalo Creek, it's hard to fathom what even a minor earthquake would do to the 225 slurry ponds that sit above old mine shafts. In Kentucky, most people have no way of even knowing where these huge ponds are—no public maps exist—and almost none of the permitted ponds have emergency plans on file, as is required by law. The worst-case scenario might involve a 250-foot-tall dam that holds back slurry at a Massey operation in Whitesville, West Virginia, right above the Marsh Fork Elementary School, where two hundred students are enrolled.

Mickey McCoy, an Inez high school teacher and local activist, picked me up outside the Inez Motel (he called it the Bates Motel), and we drove up to Coldwater Creek. McCoy wore his dark hair and beard long and untamed. He looked as though he would be more comfortable on a Harley than in his pickup, and indeed McCoy later bragged about the night his teenage daughter wowed an entire biker bar singing country covers.

We passed a series of shotgun houses, then McCoy pointed out the window to the home of a local coal magnate. It was a mansion with a French-style garden in the front and a Victorian greenhouse off to the side. "My granddaddy used to have a hog pen there," McCoy said. "I always thought it was a little better use of that space."

As a third-grader in 1964, McCoy shook Lyndon Johnson's hand when the president came to Martin County to announce his War on

Poverty. One day in the early '80s, McCoy was sitting on the courthouse steps, listening to the mayor of Inez gloat that no one had the nerve to run against him in the next election. So McCoy decided to go door-to-door, telling everyone in Inez to vote for him. And at age twenty-nine, he won as a write-in candidate. As mayor, he called for a special investigation of the state police. He took on corrupt county attorneys. As he was driving home one night, somebody ran him off the road.

But on the day I met McCoy, he was keeping it between the ditches. It was sunny, and the valley that surrounds Coldwater Creek looked verdant, remote, peaceful, much as it did the day the slurry pond broke. "We were having a local harvest festival," McCoy said. "There's probably some irony in that." Though the pond broke at three A.M., the people who lived on Coldwater Creek knew nothing about it until they saw the sludge rolling through the valley. Martin County Coal blocked off the road we were traveling to keep the media away. "But I was in good with the sheriff," McCoy said, and he brought him up here. "It looked like—" McCoy spread out his arms as if he were trying to grasp something massive, something indescribable. "The thing just enveloped the whole area. It was like someone poured a huge black milkshake out over the mountain. It just ruined everything." Today, still sixty-one other impoundment ponds sit above underground mines, and many of those ponds are not separated from the underground mines by a hundred-foot barrier, as is required by law.

McCoy veered to miss a box turtle that was crossing the road. "That's a sign it's going to rain," he said, "when the terrapins are pointed away from the creek bank."

I commented on how green the valley looked now.

"People think the slurry's gone, but it's not," McCoy said. He pulled off on the shoulder above the creek bank and grabbed a shovel from the bed of his truck, and we climbed down to the water. But before we even reached the creek, I noticed a dry black substance surrounding the young maple trees. McCoy jabbed the blade of his shovel into it and turned a clod over. "Slurry," he said. "It hasn't gone anywhere."

McCoy dug up a shovelful of sand from the bottom of the creek and turned it over on the bank. What oozed out was a viscous glob of coal sludge. I scooped some of it up in my empty coffee cup to take home. A few months before, at a Lexington forum on the Martin County disaster, Kentucky Coal Association president Bill Caylor had claimed that coal slurry was as harmless as mud, to which one angry young man in the audience countered, "Then let's go to Martin County. I'll eat some mud and you eat some slurry." Caylor replied that he would be happy to eat slurry. McCoy had been there that night, too. Now he smiled at what I was holding in my cup and said, "Take that back to Bill Caylor."

After the spill, at the request of local citizens, the federal Agency for Toxic Substances and Disease Registry (ATSDR) came to Martin County, tested the soil and water, then concluded that it posed no threat to local residents. An independent group of researchers at Eastern Kentucky University, however, using the same data as ATSDR, found that the slurry that remained in Coldwater Creek contained significantly higher-than-normal levels of cadmium, mercury, and nickel. There was 3,810 times more aluminum in the water than federal standards allow, and 6.6 times as much arsenic. ATSDR didn't even test for acrylamiae, the carcinogenic flocculent used to clean coal.

Long-term exposure to these contaminants through groundwater, drinking water, and the buried sludge that seeps to the surface during rains could certainly lead to higher cancer rates. And given that all these heavy metals bioaccumulate in the body, the university team found ATSDR's conclusion premature at best, and irresponsible at worst.

Back in the truck, McCoy started to vent about another federal agency, the EPA. "They would put hay bales across the creek as if the water was going to be filtered by that," he said. "They turned the ground over so you couldn't see the sludge. It was a real Spanky and the gang cleanup." Coldwater Creek certainly deserved to be designated a Superfund site. But under the Bush administration, the Superfund budget has been reduced by 50 percent, and its polluter-pays fees ran out in September 2003. Therefore, Martin County Coal had little trouble convincing the federal government to treat the 300-million-gallon spill as merely a violation of the Clean Water Act.

When we reached the last house in this hollow, McCoy pulled into the driveway. Glenn Cornette was sitting on the porch of his white, double-wide trailer, which stood on a small bluff. Between the creek and Cornette's house, a flat stretch of land spreads up to the head of this hollow. Cornette, who looked to be in his seventies, wore a camouflage cap and jeans. He had just returned from squirrel hunting. McCoy handed Cornette two jars of pickles he had put up. Local drinking water, pumped from the Tug River, is still so contaminated that McCoy has to buy the water he used for canning.

Cornette told me his family has been farming this eight-acre bottomland for three generations. Before the slurry came rolling through, this fertile soil was filled with corn, wheat, and potatoes. Now there is only grass and a few dying walnut trees. Four feet up from their base, the trees still bear the black stain of slurry.

Cornette showed us pictures from the day of the pond break. In them, walnuts that were shaken from the trees were bobbing in the sludge like gray Christmas ornaments. The flood of coal waste runs an inch beneath Cornette's wooden bridge and has destroyed his gas line. I looked through all the photos, each showing a different angle of the same destruction. And though they were taken with color film, they might as well be black-and-white. The slurry took all the color out of this valley and turned it into a gray apocalypse.

Martin County Coal tried to buy Cornette's property outright rather than settle a lawsuit for its damages. Cornette refused. This was the only home he knew. "You can't put a price on something that ain't for sale," he told me.

Cornette has been compensated through a class-action lawsuit, but he has nowhere to farm and no creek that his grandchildren can play in. The EPA told him they would dredge out the sludge from his bottomlands and replace it with topsoil. But Cornette showed me photos of the coal waste simply being buried under rock from the mine site above his house.

"EPA is a dirty outfit, I'll tell you," he said, smoothing his short gray hair back under his hat. "They lied to me like dogs. They told me they were going to put some topsoil back here. All they put was rock."

McCoy chimed in: "At one of the town meetings, the EPA told people that there was nothing in the slurry that wasn't on the periodic table. Well, hell, there's a whole lot of things on the periodic table that can kill you." And it is in the slurry: mercury, arsenic, cadmium, lead.

But what particularly upset local people was the chummy relationship between the EPA representatives and the industry lawyers. "The watchdogs of the environment have become the guard dogs of the coal industry," McCoy concluded. "At one of the local hearings, they sat up there on the podium with the company lawyers," McCoy recalled. "And they were all drinking bottled water. We told them, 'Go over there to the damn water fountain. Try the shit we have to drink.' None of them would."

Recently, the most notorious case of regulatory corruption has centered on the federal government's mistreatment of mining engineer Jack Spadaro. In 1972, the twenty-three-year-old Spadaro was asked to be the staff engineer for the governor's commission that investigated the Buffalo Creek disaster. He found that five separate government agencies knew the Buffalo Creek dam was unstable, but each assumed that repairing it was the responsibility of one of the other agencies. The realization that Buffalo Creek could have been avoided affected Spadaro profoundly and led him to spend the next thirty years studying impoundment dams and working for the U.S. Mine Safety and Health Administration (MSHA) to ensure better regulation of existing slurry ponds and dams. In 1996, he was named director of the Mine Safety and Health Academy, where he trained over six hundred inspectors.

After the Martin County spill, Spadaro was named the number-two man on a team sent to investigate the causes of the pond break. Then Assistant Secretary of Labor Davitt McAteer told the team, in Spadaro's words, "to find out what got mucked up down there" (well, Spadaro told me, it was a word that rhymed with "mucked"). Spadaro calls

McAteer "a safety advocate," a man who truly wanted to find the cause of the spill and make sure it didn't happen elsewhere. What the team found was disturbing. To Spadaro particularly, it looked too much like the Buffalo Creek scenario.

The investigators discovered that after a 1994 spill from that same impoundment pond had released 100 million gallons of slurry, an MSHA engineer, Larry Wilson, made nine recommendations that needed to be implemented before the impoundment pond was used again. None of the recommendations was followed, and Martin County Coal actually began filling the pond back up on *the same day* as that first spill. Then Scott Ballard, a mining engineer who had worked as a consultant for Martin County Coal, reported to MSHA that after his own investigation of the pond following the 1994 spill, Martin County Coal had only added some coarse rock around the bottom of the pond. "It was never intended to prevent a breakthrough in any form or fashion," Ballard told MSHA. "In fact, the question was asked during the review process, 'Will this prevent it?' and the answer was emphatically 'No.' There's no guarantees. There's nothing here that will prevent a breakthrough."

Who asked that question? Spadaro found that at least five Martin County Coal executives were aware of Ballard's findings, and the risk of another slurry flood, but did nothing. What's more, there was already on the mine site a filter press system that removed the water from the coal slurry and buried the remainder on-site as a solid. This method is far safer than impoundment ponds, but because it cost $1 more per ton of coal, Martin County Coal abandoned it in the '90s and went back to filling up sludge ponds. In addition, Spadaro and the other investigators bored and extracted samples from below that pond and found that the wall between it and the open mine shaft was not one hundred and fifty feet, as Martin County Coal claimed, but less than eighteen. Martin County Coal had falsified its mine maps to make the wall look thicker.

By the end of 2000, Spadaro and the other investigators thought they had collected enough evidence to charge Massey Energy, the parent company of Martin County Coal, with willful and criminal negligence.

And then George W. Bush was elected president for the first time.

It is no secret, and no surprise, that Bush, along with Kentucky senator Mitch McConnell, received millions of dollars in campaign contributions from the coal industry. Massey Energy alone donated $100,000 to a Republican Senate Campaign Committee headed by McConnell. I mention the Kentucky senator because, aside from being the Senate's lead opponent of campaign finance reform, he is also the husband of Labor Secretary Elaine Chao—to whom MSHA answers.

Chao appointed Andrew Rajec, a former McConnell staffer, to the internal review team put together to investigate the allegations regarding MSHA negligence. But Rajec had no mining experience, and thus his apparent job was to protect Massey Energy and report to Chao and McConnell any of the team's preliminary findings.

Within days of Bush's inauguration, Dave Lauriski replaced McAteer as assistant secretary of labor. Lauriski was the general manager of Energy West Mining before joining the MSHA. Reflecting on his tenure as assistant secretary three years later, Lauriski told the Oklahoma City *Journal Record*, "The industry has always been good to me. I just hope that I've given back as much as I've received." Indeed he did. During his tenure, he worked to lower coal dust standards that protect workers from black lung, hand out no-bid contracts to friends, and transfer MSHA officers who levied fines that Lauriski's mining friends thought were too harsh. The rap sheet on Lauriski is long. Lauriski's daily schedule, obtained through the Freedom of Information Act, shows a repeated pattern of behavior whereby Lauriski met with high-ranking coal operators and lobbyists right before he moved to weaken another regulation standard. But his conduct concerning the Martin County spill was particularly egregious.

After taking over as assistant secretary, Lauriski named a new team leader, Tim Thompson, to head the Martin County investigation. Thompson told Spadaro and the other investigators to wrap their work up immediately. The investigators wanted to cite Martin County Coal for eight violations, including willful negligence. Thompson had that reduced to two, with a total fine of $110,000 (Massey appealed that negligible sum, and the fine was reduced to $5,500). Spadaro refused to sign the report. Later, Thompson called four of the other investigators into his office and asked them to sign off on the report without reading it; he gave them only the signature page. The men objected, and Lauriski agreed to let them see what they were signing. Spadaro, however, still refused.

Then one day Lauriski called Spadaro at his Beckley, West Virginia, office, and said he really needed Spadaro to sign the report—that he, Lauriski, was in a "tight spot." Who had put him in that spot? According to Spadaro, it was Mitch McConnell. "He is the most corrupt politician I've ever come across," Spadaro told me at his home in Hamlin, West Virginia, in January 2005. And McConnell, his wife, and his wife's appointee, Lauriski, were doing whatever necessary to protect Massey Energy.

Spadaro told Lauriski that under the circumstances, the best thing he could do was simply take Spadaro's name off the report, which Lauriski did. But that wasn't the end of it. On May 5, Lauriski met, as he frequently did, with Massey Energy CEO Don Blankenship. Exactly a month later, Lauriski sent Spadaro to Washington, presumably on MSHA business. While Spadaro was gone, federal officials searched his Beckley office and changed the locks. Then he was abruptly placed on administrative leave of his duties at the Mine Health and Safety Academy. As justification, Lauriski accused Spadaro of "abusing his authority" while superintendent of the academy. As it turned out, this abuse amounted to giving free room and board to a field officer who had developed multiple sclerosis.

And all of this was done in an effort to protect Massey Energy from criminal prosecution, at the expense of hundreds of people who live below that impoundment pond in Inez, Kentucky. The logic is easy to follow: Massey Energy contributed heavily to George W. Bush's presidential campaign; Bush won the election by substantially outspending Al Gore; Republicans always stand to lose with campaign finance reform; therefore, President Bush rewarded Mitch McConnell for his decade-long fight against campaign finance reform by naming McConnell's wife as labor secretary; and finally, Elaine Chao made sure that her underlings at MSHA protected a major Republican campaign contributor, Massey Energy. It's a pattern of corruption that neatly flows full circle. Everyone has everyone else's back, and the one whistleblower who tried to speak out for the public's interest is left spinning in the wind.

When the bogus charges that Spadaro had abused his authority didn't pan out for Lauriski, he transferred Spadaro to an office in Pittsburgh, far away from his family in West Virginia. On July 4, 2003, Spadaro filed a complaint with the U.S. Office of Special Counsel, claiming that the charges made against him by MSHA were retaliatory. But it turned out that even this special counsel had only found in favor of one whistleblower, and there was no guarantee that Spadaro would ever get his job back at the Mine Health and Safety Academy. So in the end, he agreed to what he characterized as a "no-fault divorce" with MSHA. He would drop charges, MSHA would restore him to his original, higher pay grade, and Spadaro would retire back to his mountain home high above Hamlin.

But this retreat, where we sat by a soapstone stove rehashing it all, looked to me only like a short respite in a longer battle. Spadaro intends to keep fighting from the outside by helping coalfield residents bring litigation against companies that are destroying their homes, land, and water. The federal government certainly can't be trusted, and Spadaro said that both state and federal regulatory agencies "have failed miserably to enforce the Surface Mining Control and Reclamation Act."

He put another log in the stove, then said, "We have to admit that a handful of coal operators are getting rich at the expense of everybody else. You can't just fill every headwater stream in Appalachia and think there won't be problems. It's insanity. We have to realize that there's a limit. The ecosystem can't handle it."

I looked out the window at Spadaro's own 125-acre ecosystem and imagined how it would look in the spring, when wildflowers and trees bloomed throughout the understory. The steep ridges looked pristine and unscathed. They looked the way a mountain should look.

Mickey McCoy and I left Glenn Cornette to start skinning the squirrels he had bagged that morning. We drove around for a while and looked at some more strip jobs and valley fills. He told me one story after another about his run-ins with the local coal moguls. Finally McCoy, who I was starting to like immensely, said he couldn't bear the thought of me sitting by myself another night at the Bates Motel. He kindly invited me to drop by his house later that evening for a drink.

When I arrived, he was "straightening up" his garage, which he had converted into a kind of bar/museum/storage area. Motorcycle boots and an iron kettle hung from the trusses overhead. Left-wing posters and bumper stickers were stuck in seemingly random places. Various commemorative bottles of Maker's Mark bourbon, each dipped in a different color wax, stood behind the bar, guarded by a Kentucky Wildcat statue wearing Mardi Gras beads.

McCoy took up a position behind the bar and started packing ice into a couple of tumblers. He poured me a dram of his favorite bourbon, a small-batch product called Kentucky Gold, and told me about the musical and theatrical accomplishments of his two teenagers. Soon his wife, Nina, came in with a pizza. She teaches biology at the same local high school where Mickey teaches English. And since the slurry rolled pass their house four years ago, she has fought for answers and accountability just as hard as her husband. In fact, one gets the idea that the two are seldom apart. They complete each other's thoughts and speak with the same flair, the same righteous anger.

Nina remembers 1988, when some representatives of the National Coal Council came to Martin County. They took many of the local science teachers to Jenny Wiley State Park for a lavish dinner and some laudatory speeches about the contributions eastern Kentucky had made to the nation's well-being. Then the coal lobbyists dropped the hammer. "They started talking about how there was no such thing as global warming," Nina remembered. "I was livid. I stood up and just started shouting." She smiled. "I think that's when I got on their bad side."

The McCoys have always been a public-spirited couple—Mickey brought a sewer system to Inez when he was mayor—but it wasn't until the 2000 slurry spill that they decided to fight the coal industry head-on. "That's when we kicked it into high gear," Mickey said. Now the McCoys are five years from retirement, and looking forward to the day they can fight strip mining full-time. Though Mickey concedes that he has isolated himself from his more conservative neighbors, he plans to stay and fight the coal companies. "I'm not going to give them the damn satisfaction of seeing me move," he declared, pounding the bar.

When it comes to the subject of coal, I had already learned, Mickey becomes animated quickly. "It's like this land is being sacrificed for the sake of this idea that we have to have cheap coal," he said, holding out his arms as if waiting for some explanation that was more rational. "It's idiotic, because we have a sun ball up there that generates all kinds of energy and we have the capability of getting it. But the oilmen and the coal men don't have the capability of owning the sun. They can't regulate it. In that sense, alternative energy is not in the capitalist logic. It's harder to own the wind and the sun. Hell, it shines everywhere. It's too, too . . . *democratic*."

But when it comes to alternative energy, or any other change that would place limits on the coal industry, the problem is getting there from here. The people who have been hurt the worst—the Martin County residents who live both below and downstream from these enormous mountaintop-removal sites—are the same people who feel the most powerless. Inez's drinking water is pumped from the Tug River, where aquatic life was annihilated by slurry. Nina said that last summer she was on the phone with someone from the EPA. "She told me the Tug River was so degraded and had been for so long, they had stopped monitoring it," Nina said. "Nobody here drinks the water, but nobody does something about it. Too many people are waiting for Erin Brockovich to come in here and fix it for them. Well, guess what?" Nina was now waving one finger in the air. "Erin Brockovich isn't coming. We have to rise up and say, 'Give us some clean water.' Then something will happen."

Now she wants to get the EPA's attention—get *somebody's* attention—and get the Tug River cleaned up. To that end, she has started SAVE: Save Appalachia's Vital Environment. Her first goal is to organize enough people to make enough noise.

As for Mickey, his immediate plan of action is to put a huge sign in their front yard that will read:

GOD WAS WRONG

SUPPORT MOUNTAINTOP REMOVAL

That should get a rise. "People always tell me I'm going to get into trouble," Mickey said, raising his eyebrows in feigned fear. "They say, 'Oh, don't do that, you'll get in trouble.' Or, 'Oh, don't say that about him, you'll get in trouble.'" Mickey set his tumbler down with conviction. "I say, What kind of trouble?"

The question hangs in the air for a moment. And then the answer becomes clear: the right kind.

June 2004
LOST MOUNTAIN

I'm riding in the back of a white Fish and Wildlife pickup, trying to keep my breakfast down while the driver, a guy named Charlie, is kicking up large clouds of dust as we speed across an abandoned strip mine. An elk calf is the reason for our haste. All summer, a team of wildlife biologists has been trying to catch and collar these young elk so they can measure the animals' survival rate and predict population growth rates in eastern Kentucky. With weaker immune systems than older elk, these calves are particularly susceptible to the meningeal worm parasite.

This morning, a helicopter is circling overhead. In it, John Cox is tracking one calf with an infrared scope and radioing its whereabouts down to Charlie. Once we get close enough to the calf, Charlie hits the brakes and the doors of the truck fly open. Charlie and two other technicians go charging after the calf through the high lespedeza. I also leap into a kind of action and start chugging off in the same direction. The disturbed ground is extremely uneven, and every footfall looks to me like an ankle about to snap. The idea is to close in on the calf from all sides; its mother, by this time, is watching frantically but helplessly from a distance. Then, whichever direction the calf bolts, the closest person tries to tackle these gangly forty pounds of legs and hooves while someone else gets a radio collar around its neck. "Release your inner predator," someone told me earlier. Well, I was trying.

When I caught up with the others, this particular calf had sought cover in a small grove of exotic olive trees. I took up my position in the circle and crouched down like a third baseman waiting for a line drive. I stood about fifteen feet from the calf, close enough to observe the white markings on its silky brown torso and the nervous look in one eye. Fortunately, it didn't shoot off in my direction, and when the closest technician lunged after the calf, he missed his tackle. The last we saw, the young elk was bounding over a hill, back into forest cover. John radioed down that the helicopter was low on fuel and heading back to the airport. The day was done. We would all rendezvous at a local diner called Chaney's.

Earlier in the morning, we had seen a group of bull elk browsing on sumac leaves near the edge of the mine site. They are regal animals, with their handsome antlers and oily black necks. In the fall, when elk are breeding, the bulls' mating call is a "bugle." What it sounds like is a very distressed saxophone, something from the repertoire of Albert Ayler. In October, the bull elk will corral a stable of cows, and if another bull moves in and isn't up to a fight, he may get badly gored by one of those antlers. In the summer, however, all is forgiven, and the bulls, who want nothing to do with child rearing, form small bachelor parties that stretch across fourteen Kentucky counties.

The elk (*Cervus elaphus*) was part of a rich array of large mammals that roamed the eastern United States during the Pleistocene era. Forty of those species went extinct 10,000 to 15,000 years ago, and the last Eastern elk was finally shot by a hunter named Jim Jacobs in 1867. In the late '90s, 1,500 Western elk were reintroduced into eastern Kentucky. There are important ecological reasons to bring the elk back to Appalachia. According to Dave Maehr, a conservation biologist who played a vital part in the elk reintroduction, the elk reestablish evolutionary relationships that are native to the region but have been disrupted for almost two centuries. The elk keep overgrazing deer in check and add "community complexity" to the local ecosystem. In other words, the elk bring back the biome's earlier biological diversity and balance.

Anyone who believes, as I do, that natural selection has organized ecosystems into a complex, symbiotic relationship of species—and that *nature knows best what it is doing*—would commend Maehr's work with the reintroduction of elk into Appalachia. Unfortunately, the elk project has also been appropriated by the coal industry in troubling ways. In a pamphlet called *Pocket Guide to Kentucky Coal Facts* put out by the industry, one graphic shows cartoon elk bounding happily over a mountaintop-removal site. The pamphlet boasts that mountaintop mining has provided an ideal home for "free-ranging" elk.

Some coal operators have started shuttling groups up to mine sites at dawn so they can watch the bull elk grazing in the early fog. What they have created, the companies will tell you, is excellent wildlife habitat. But a rolling grassland is hardly ideal for elk. David Ledford of the Rocky Mountain Elk Foundation calls most reclaimed land in central Appalachia "pretty inadequate," and he estimates that only 35 percent of reclaimed land can support fish and wildlife. This year, the Rocky Mountain Elk Foundation held a "Mine Reclamation for Wildlife" summit in Louisville in an effort to encourage coal operators to create better wildlife habitats. At that meeting, Jon Gassett, the interim commissioner for the Kentucky Department of Fish and Wildlife, pointed out the obvious: "Coal companies are not going to do good wildlife management on reclaimed land if it costs them anything." Yet one would hope that as the elk population grows in Kentucky—it has reached 5,300—hunters, a traditionally conservative group, might begin to pressure the industry to provide better hunting grounds. This year, some 16,000 hunters applied through a lottery for a total of 100 permits to hunt the elk. Better reclamation would obviously mean more elk and more permits to hunt them.

And the support of the hook-and-bullet crowd would give environmentalists a welcome if unlikely ally. Hunters, for one thing, are not as easy to marginalize in the media. They usually don't have long hair or wear tie-dye. And along with wanting better habitats for game animals, I think many sportsmen, such as the Baptist Bow Hunters group I met briefly last year, would respond positively to Steve Peake's position that the natural world is God-given and deserving of our stewardship. In the current political climate, one thinks of environmentalists hailing from blue states and hunters from red ones, but for a campaign against mountaintop removal to succeed, it will, like the civil rights movement, have to transcend right-left divisions and be about something far more fundamental than politics.

At Chaney's, the guys from Fish and Wildlife are telling me one rapid-fire elk story after another. There was the time an elk that didn't want to be led into a trailer kicked Charlie in the head, and the other time one rose up on its hind legs and boxed him like Lennox Lewis. And then there's always some hapless soul who doesn't know nearly as much about the animal in question as the storyteller, someone who, for instance, tried to use a cattle prod on an elk and barely lived to tell about it. He's the foil, the guy you shake your head at in supreme confidence that you could never be him. Most Fish and Wildlife people, I've noticed, are very good storytellers. They are hunters, after all—even if the hunt has been reduced to only collaring prey—and part of hunting is the story of the hunt, told around the primordial campfire. Or in our case, a large wooden table covered with plates of sausage and eggs.

The food at Chaney's is good and ridiculously cheap. While we are eating, my gaze drifts along one wall. At one side hangs a cluster of family photos, mostly children and grandchildren. Beside those is a large painting of a bull elk with an impressive rack of antlers. And then, to the right of the elk is a 24x30-inch photo of a detonation on a strip mine, caught just as flyrock fills the air. As I work on my biscuits and gravy, I try to make sense of this iconography. In what way is the photo of a strip mine similar to a family portrait, or an image of native wildlife? The images are all framed impartially, handsomely; they all represent points of pride.

Someone in the family probably works for, or perhaps owns, a strip mine. One reason that more people in eastern Kentucky don't speak out against mountaintop removal is that nearly everyone has a family member, or an extended-family member, who works on a strip mine.

To make known one's opposition to the practice is to complicate Sunday dinner, as in Gurney Norman's short story "Home for the Weekend," from his collection *Kinfolks*. In it, Wilgus Collier, a college student from eastern Kentucky, returns to his grandmother's for Memorial Day Weekend, where the Collier clan has gathered, and where Wilgus informs his uncle L.C. that "strip mining was the work of the devil and that the whole coal industry ought to be nationalized."

> "Why you goddamn . . . *beatnik*," L.C. snarled, just as Junior's wife Betty, who'd been shouting at Delmer, turned toward Wilgus and said, "For your information, Wilgus, I have a brother working at a strip mine."
>
> Wilgus started to ask what that had to do with it, but L.C. shoved his finger in his face and said, "Why Wilgus, where do you think jobs around here would come from if it wasn't for strip mining? People has to *work*."
>
> "By God, L.C.," Delmer [Wilgus's uncle] shouted from across the room. "They'd come from deep-mined coal like they always done. There'd be work aplenty if they'd stick to deep mines."
>
> "Delmer, that's all well and good," said Junior. "But the thing is, they never would of started this strip mining if John L. Lewis had any guts."
>
> Delmer choked. He grabbed somebody's ice tea off the table, downed half of it, and when he got his voice back he looked Junior in the eye and said, *"What did you say?"*
>
> "I said if John L. hadn't let us down, the miners would of taken the whole thing over by 1950, and *working* people would call the tune."
>
> "By God," Delmer shouted, "if the union had stuck together . . ."

L. C. waved his arms in the air and said, "The Catholics and the Jews!"

In 1950, United Mine Workers of America leader John L. Lewis agreed that if coal companies provided their miners with heath care benefits, the UMW would not fight against any further automation within the coal industry. Many people, including the fictional Delmer Collier, thought that was the beginning of decreasing jobs and increased stripping, using dozers, rock drills, and the kinds of explosives depicted in the photo on the wall at Chaney's. But nostalgia about the days of strong unions and hard, honest work underground still remains throughout Appalachia. And many people make no distinction between deep mining and strip mining. Here at Chaney's, the image of a strip mine—incongruous to me beside family photos—is simply an emblem of success, family pride. Whether the Chaney family has its own Wilgus, or whether it has the same kind of fights as the fictional Colliers, I, of course, have no way of knowing. But what I do know, at least what this photo reminds me, is that nothing in eastern Kentucky is ever as simple as it may appear to an outsider.

After breakfast, I leave the Fish and Wildlife guys to their work and head out for my monthly inspection of Lost Mountain. The road leads through a small community called Rowdy, which supposedly takes its name from the disposition of its residents. Its post office is the size of a suburban toolshed. I have not seen much evidence of disquiet on my drives through Rowdy, though I do pass a narrow strip of valley where one man is raising an impressive flock of red roosters (in a few months from now, a few counties away, a major cockfighting venue will be raided and shut down). I also pass the home of Silas and Edith Miller. A few years ago, the couple had agreed to let Horizon Natural Resources, the parent company of Leslie Resources, mine the fifty-acre mountaintop behind their home in exchange for $2 for every ton of coal hauled out. Silas Miller has black lung and needs the money to pay medical bills. But now Horizon has filed for bankruptcy and says it is unable to pay the Millers the $688,672 it owes them.

I take a right onto the road that leads over Rowdy Mountain. Or at least what's left of it. About a hundred feet above the houses and trailers sits a silt pond, meant to catch heavy rains and sedimentation. It is dammed on the side facing the valley, and a steady stream of water trickles out of a culvert near its rim. The sides of Rowdy Mountain rise above the pond, gray and absolutely bald. Vertical bulldozer tracks are still visible from where all of this loosened rock was compacted. There is certainly nothing here to hold back rain or stop erosion. Only sparse patches of grass struggle to take hold in the dozer tracks. But a few hundred feet higher, it becomes quite clear that whoever mined this mountaintop soon abandoned any notions of reclamation. Where the road crests, I look out over a bench piled with unshaped mounds of black and gray spoil. It looks as if someone detonated 500,000 pounds of explosives and simply drove off, never to return.

There are two kinds of abandoned mines. The first are those that were abandoned before SMCRA was passed in 1977. Under the Abandoned Mine Lands Act, coal operators are required to pay a surcharge on every ton of mined coal—35 cents for stripped coal and 15 cents for coal mined underground. That money goes into the AML fund for reclamation of pre-1977 sites. But when coal use decreased in the '80s and '90s, the fund nearly dried up. Today, only "priority" cases—

mines that pose imminent risk to people living below them—get funded for reclamation.

The second type of abandoned mine sites involve a more contemporary problem. All coal operators have to put up bond money before they are granted a permit to mine. The bond is meant to cover the cost of reclamation should the state, for whatever reason, have to step in and complete the job. When mining companies themselves have completed a portion of the reclamation approved for a particular site, such as hydroseeding the lower sides of a mountain, then part of their bond is released back to them. But as the earlier case of the Dean brothers in Harlan County demonstrates, the actual job of reclamation usually costs more than the posted bond. So coal operators will often simply forfeit their bond and leave the site in question unreclaimed. A 2004 report from the federal Office of Surface Mining (OSM) conducted a follow-up evaluation of unreclaimed sites from 1997; it found that in 60 percent of the cases, the bond money was not adequate to complete the reclamation. An OSM field officer told me that the number is more like 100 percent. So who foots the bill for the rest of the reclamation process? No one. It just doesn't get done.

That seems to be the case here at Rowdy Mountain, and it compounds the farce of reclamation with the fact that very little of it is being done at all. Of the 5,825,756 acres under permit to be mined in 2002, only 73,407 were released from Phase 3 bond, which is complete reclamation. In 2003, according to OSM's most recent oversight review of Kentucky, 89 strip mine sites forfeited their reclamation bonds, and 88 of those sites went unreclaimed.

The road forks at the bottom of Rowdy Mountain, and I turn right toward Harveytown, where I park my truck and start up the back side of Lost Mountain. In April, acting on a citizen's complaint, the Kentucky Department of Natural Resources found that the berm circling the northwest side of the Lost Mountain mine site had no erosion control, and furthermore, drainage was passing off the mine site without proper sediment control. The environmental inspector from Natural Resources ordered the company to seed and mulch the mounded rim. Today, grass and staghorn sumacs are slowly taking hold on the berm, and a silt pond has been dug at the western edge of the site to filter out sedimentation before the rainfall drains down toward the community below. Recent heavy rains have carved totemic-looking figures along the black faces of the coal seams. Stagnant gray water puddles in the pits all along the number 10 seam. Much of the overburden has been piled in the middle of the site, through which graders have leveled a makeshift haul road to truck the coal out. Inside the outer berm, the mine site looks something like a geological tidal pool, where waves of rock and earth are constantly shifted back and forth, carved into temporary crests that will fall again into a deep hollow once its coal has been removed. Of course, in a real sense, these Cumberland mountains *are* waves, alluvial currents shifting over eons but measurable only in what geologists call "deep time." They are moving so slowly that the earth feels solid beneath our feet. And when they move in natural time, they take their trees and topsoil with them. But strip mining has sped up geological time to the point where mountains that took 290 million years to form can be leveled in a matter of months, and their coal consumed in even less time.

The story of the last hundred years is the story of acceleration. In 1905, while the brothers Wright were guiding their first flyer around Kitty Hawk, Henry Adams wrote his essay "The Law of Acceleration." There he hypothesized that human civilization was careering into the future with an ever-increasing speed that we are powerless to stop. The speed by which technological society advances could be measured by squaring the achievements of the last generation—that was Adams's formula, that was his *law of acceleration*. His gauge? The consumption of coal. "The coal-output of the world, speaking roughly, doubled every ten years between 1800 and 1900," he wrote. In the last forty years, coal production has jumped from 5 million to 12 million short tons. There was, Adams admitted, a chaotic upsurge, a "vertiginous violence" associated with such acceleration, but he believed the modern mind could harness that force and use it for good. "At the rate of progress since 1800," wrote Adams, "every American who lived into the year 2000 would know how to control unlimited power. He would think in complexities unimaginable to an earlier mind." But alas, we don't, and we don't.

THE ECOVILLAGE

The lush bluegrass region of Kentucky comes to an abrupt halt at the Kentucky River, where spectacular walls of limestone drop two hundred feet straight down to the water. The Camp Nelson Veterans Cemetery sits right at the edge of these limestone palisades, where its uniform white headstones offer a commentary on the contrast between human mortality and the deep geological time to which these rock walls bear witness.

During the Civil War, Camp Nelson was a Union stronghold precisely because these steep ramparts rendered Confederate troops powerless from the other side of the river. As a training outfit, Camp Nelson produced two of every five black Union soldiers. Since the Emancipation Proclamation only freed Southern slaves, Camp Nelson had no trouble enlisting local black men who joined the army to gain their freedom. The Union army showed them its gratitude by driving four hundred of the recruits' wives and children from the camp on a frigid November night in 1864. One hundred and two of them froze to death on their trek to Nicholasville, the closest town. Joseph Miller was one soldier whose wife and child were expelled that night. Miller slipped out of camp and walked the eight miles to Nicholasville, where he found his wife holding their dead son in her arms. The next night, Miller returned to dig a grave to bury his child.

What does this sad story have to do with strip mining? After the terrible purge at Camp Nelson, an abolitionist minister—a saint, really—named John Fee arrived at Camp Nelson to set up a church for black soldiers. And after the war ended, Fee bought land in Madison County, where he started Berea Literary Institute, the first college in the country to accept black students. Today, Berea College devotes itself almost exclusively to educating the children of lower-income Appalachian families. In 1999, Berea president Larry Shinn set up a Subcommittee on Sustainability (SOS) to explore ways to make the college less dependent on coal. Out of SOS came the decision to set up a minor degree program in Sustainability and Environmental Studies (SENS). Ecologist Richard Olson was hired to direct SENS, and two years later, when the college decided to expand its single-parent housing, President Shinn directed Olson and his students to develop a housing community that would be environmentally sustainable and that would push forward the college's historically progressive philosophy. As part of that mission, Berea sets aside one housing complex, complete with a daycare facility, that is devoted to single parents. The result is a collective of apartments, gardens, and greenhouses called the Ecovillage.

In his most recent book, *The Last Refuge,* David W. Orr writes of his colleagues in the environmental movement, "The public, I think, knows what we are against, but not what we are for. There are many things that should be stopped, but what should be started?" I drove down to Berea's Ecovillage to get some idea.

I met up with ecologist Richard Olson in the Ecomachine, a greenhouse lined with a large septic tank and a series of open tanks filled with calla lilies and other large-leaf tropical plants. Olson is a lean man with graying curly hair. He wore jeans and a sweatshirt. One of his assistants was moving between the tanks, carrying oddly shaped test tubes. Here, all of the waste from Ecovillage's toilets is slowly transformed, using natural processes, back into swimmable water. "All of the processes that take place here are biological processes," Olson told me. "There's nothing magic about it."

What does look somewhat magical is the septic tank, painted by a student in bright, rain-forest colors. From there, the waste is transferred to the plant-filled, open tanks, where it is converted to ammonia, then nitrogen, and finally into nitrates. "It works," Olson said, "because what we consider waste—bacteria—the plants think is food." The plant roots provide surface area for the bacteria as the water moves from tank to tank, becoming increasingly cleaner. Finally it flows out of the Ecomachine to a subsurface wetland—nobody likes mosquitoes—where it undergoes a denitrification process. Then the clean water is pumped back into the apartments for clothes washing and toilet flushing. In this way the Ecomachine, as I would come to see, embodies the main principle of the entire Ecovillage: follow the laws of nature. If a human economy creates massive amounts of toxic waste and a natural economy creates none, perhaps we should pay better attention to what the latter is doing.

Olson led me out to the open courtyard area of the Ecovillage. Ten feet beneath that lawn run pipes that circulate water and antifreeze. Because that subsoil always maintains a 57-degree temperature, Olson explained, the Ecovillage uses a ground-based heat pump that exchanges off that median temperature instead of 90 degrees in the summer and 20 degrees in the winter.

Six of the new apartment buildings, housing fifty separate units, line both sides of the rectangular lawn. Most of their windows face south to maximize the use of passive solar heat. Because of a built-in thermostat, each window will automatically open in the summer to vent out excess heat. The exterior of the apartments is a combination of red brick and Hardy Plank, a cement-based siding that, unlike vinyl, generates no dioxin or vinyl chloride as by-products. The walls are tightly insulated, and over each patio, a wooden trellis, covered with deciduous vines, provides shade in the summer. Inside each apartment, colored concrete floors trap solar heat during the day and release it at night. There are plenty of fans, compact fluorescent lightbulbs, a low-flush toilet, and a front-loading washer. There are no dryers. Each student hangs his or her laundry on one of those revolving clotheslines that look like an umbrella turned inside-out by the wind. What it all adds up to is a residential community that uses 75 percent less energy and water than the average American home or neighborhood.

But do students like living here? Olson is quick to point out that the Ecovillage is not run by "environmental Nazis." When a student moves in the first year, he or she is given a list of sustainable activities, such as recycling, carpooling, composting, or gardening, and is asked to perform two of those tasks. In the second year, a third task is added. And so on. But to convince students with no strong concern about the environment to live in a manner that cuts energy use by three-fourths is, said Olson, "one of the challenges we face and one of the opportunities." He went on, "To meet performance goals, architecture will only take you so far. The students who live here are not here because they have any interest in sustainability. They are here because they have kids. Our hope is that by living here they will start to pick up on some of these things and so will their kids."

Like John Dewey's Chicago Laboratory School, where children used the simple task of cooking to understand math, chemistry, and biology, the Ecovillage is an ambitious example of experiential learning. Olson hopes that knowledge about sustainability will lead to a change in attitudes, and that in turn will lead to a change in behavior. That this evolution has already begun is evident in the vegetable gardens that have sprung up in the small square plots in front of each apartment. With no gardening experience, many of the Ecovillage residents have begun to grow some of their own food.

And indeed, one can discern a closed loop of energy flow that begins with the garden and moves through the low-flush toilets, into the Ecomachine, out to the subsurface wetland, and back to the garden hose. It's one example of how, of why, experiential learning works. "Most people don't see these connections," Olson said. "We're trying to make things as visible as possible. You ask the average person in town, when they flush their toilet, where does it go, they have no idea.

These people know that when they flush their toilet it goes there"—Olsen pointed to the greenhouse.

Perhaps the most obvious achievement so far of this hands-on pedagogy is the white house that stands beside the Ecomachine. In January 2001, Olson asked the students in his Ecological Design class to research and design a house that would be completely independent of the coal-fired energy grid. The result is this two-story residence with a composting toilet, a solar hot-water heater mounted on the roof, a gutter system that collects all the water needed in the house, and a freestanding unit of photovoltaic panels that tracks the sun across the sky. "The students designed, built, and lived in the house," Olson said. "That is about as classic an example of experiential learning as you can get."

As Olson led me around the Ecovillage, a breeze picked up, and I asked if wind power was an option here. He said it wasn't. In Nebraska, where Olson earned his Ph.D., there was enough wind to provide power for the entire state. In fact, Olson told me, with the proper infrastructure, the eleven Plains states could meet the energy needs of the entire country by harnessing the wind. Instead, for now, the eleven states with coal must bear that burden.

Yet even though the Ecovillage seems to operate on principles that are antithetical to a coal-powered economy, Olson doesn't blame the coal industry. He blames Berea College, he blames me, he blames himself. "Who is destroying the mountains of eastern Kentucky and West Virginia?" he asked rhetorically. "It isn't the coal companies. It's us. I wish I could fly every student over eastern Kentucky and say, 'You did this. Okay, forget the guilt. How can we change that?' They have to understand the impacts and they have to understand the alternatives."

Over the last nine months on Lost Mountain, I had seen the impacts and now was standing in the middle of the alternative. A pretty student approached Olson with a handful of vegetables and announced that lunch was ready. Olson cast a long, admiring gaze around the grounds, then turned to me and said, "It doesn't look like such a harsh way to live, does it?"

July 2004

LOST MOUNTAIN

Patsy Carter grew up in an Appalachian hollow where the mountains of eastern Kentucky rose up in front of her house and the mountains of West Virginia fell away in back. It was a beautiful but isolated river valley; no roads reached it. All the families who lived along this stretch of the Tug Fork River, which separates Kentucky and West Virginia, had to cross the river in a rowboat, then walk two miles of bottomland to reach the roads and towns of Pike County, Kentucky, or Williamson, West Virginia.

So in the early 1960s, Carter's father and some other men from this community in Martin County started building their own road, using mattocks, shovels, and dynamite. As part of the migration of Northern journalists to Appalachia, spurred by Caudill's *Night Comes to the Cumberlands,* Chet Huntley and David Brinkley of NBC arrived in Martin County in 1965 to report on the progress of the road Patsy

Carter's father and his neighbors were building. After *The Huntley-Brinkley Report* aired its story in 1966, small amounts of money started trickling in from viewers across the nation. Five dollars from a man in Oregon bought a case of explosives. Gradually there was enough money for a small bulldozer. Slowly, the four-mile road got built, and when it was finished, the community of less than two hundred decided to name it the Huntley-Brinkley Road.

Today, forty years later, this narrow two-lane is one of the most dangerous roads in America. All day and all night, illegally overloaded coal trucks fly around its curves at speeds well over the limit. One only has to tail a truck for several of these winding miles to realize that most drivers spend almost as much time in the wrong lane as the right one. The weight limit for the road is 44,000 pounds, but drivers almost always haul more than 100,000 pounds of coal; some haul around 200,000 pounds. As a result, the pavement is badly cracked and pocked. In some places, entire sections of the road have fallen down into the Tug Fork. On federal highways, coal trucks cannot legally carry more than 120,000 pounds. But a study by the Kentucky Transportation Cabinet found that the average weight of a coal truck on US 23 was 155,000 pounds. A 1999 study found that 88 percent of commercial trucks traveling eastern Kentucky roads were illegally overweight.*

---

*Weight is not the only problem associated with coal trucks. These eighteen-wheelers also generate an incredible amount of mud and dust. When it is wet on the mine sites, mud often collects between the wheels of the coal trucks. And when the drivers head back to the roads below, they splatter those winding two-lanes with mud from the site. One hundred coal trucks a day thundering down one road can create what looks like a mudslide. Yet when residents of Ary, Kentucky, complained about the dangerous mud that coated a road traveled by school buses, the Department of Natural Resources told them that public roads do not fall under the state's jurisdiction.

In the spring of 2000, Patsy Carter's daughter Darlies was preparing to graduate from college. The two were very close; they canned vegetables and shopped for Appalachian antiques together. As Darlies was leaving work one night to come home, she called her mother to let her know she was on her way. When Darlies wasn't home in twenty minutes, Patsy Carter started to worry. When Darlies wasn't home in forty minutes, her mother got in her car and headed west along the Huntley-Brinkley Road.

Today, Carter struggles to describe what she came upon. Much of it she has blocked from her memory. But what happened that night is this: Matthew Casey, a repeat felon, had taken thirty Xanax and had been driving his overloaded coal truck for eighteen hours straight when he lost control of it and drove head-on into Darlies Carter, killing her.

"I was getting ready for a graduation and instead had to go to a funeral," her mother told me.

As for Matthew Casey, his truck was found to be fifty tons over the legal limit (on this particular rural road, the limit was twenty tons). Still, he lied and said Darlies had crossed into his lane. Joseph Meadows, the coal truck driver who was following Casey, confirmed that account. This is a tactic that both the drivers and the coal companies have perfected: They send trucks out two at a time so that if there is an accident, one driver can corroborate the other's story that he wasn't at fault. It was only after Casey was brought before a county judge that he admitted it was he who had swerved and hit Darlies's car. When the judge asked Casey if his employer, Massey Energy of Richmond, Virginia, knew he was hauling seventy tons of coal, Casey said yes. He also admitted to using cocaine and marijuana daily. But curiously, the county lost his blood work from that fatal night, and Casey ended up serving only one year in the Southern Regional Jail.

But this was far from an isolated incident. From 2000 to 2004, in Kentucky alone, over five hundred people were injured in accidents involving coal trucks, and fifty-three were killed. While Matthew Casey was still in stir, Patsy Carter's nephew, eighteen-year-old Bobby Duane Crum, who lived across the road, was hit and killed by another overloaded truck as he was driving home from school. A short time later, another coal truck veered around the mail carrier, causing its tail end to knock Crum's father, who was retrieving his mail, into a deep gully beside the road. The driver didn't even stop.

"We live in terror," said Carter's sister, Judy Maynard, whose nephew was killed by a coal truck in 1990. "It's like Vietnam around here just trying to go to the grocery and back."

But to complain about any of this is to risk retaliation. Since Darlies was killed, her mother and aunt have fought for better regulation of weight limits and drug testing for the drivers. They even blocked the road one day last summer to draw attention to their cause. The result? "I've had twenty-two flat tires in two years," Maynard said. At night as the trucks rumble by, she can hear nails hitting her mailbox. "I bought a big magnet and hooked it on a string," she said. "I have to drag my driveway with it every morning."

"Bloody Mingo County," the site of the 1920 Matewan massacre and the violent union wars that followed, sits just across the river. And in many ways, the intimidation that Mingo County coal operators have long been known for still exists, here and across the coalfields.

A year after Bobby Duane Crum's death, his mother April died at age forty-seven. "That woman grieved herself to death," Maynard said. "She prayed for the day she would die."

And Patsy Carter, more than anyone, understands. Since her daughter's death, she has had a nervous breakdown and constant nightmares, and she still finds herself doing Darlies's laundry out of compulsion. She still walks into her daughter's bedroom and thinks of the clothes Darlies had laid out to wear the next day.

"I'm a dead woman walking," Carter said as we stood on the back porch of her house, which sits high above the Tug Fork. "But I will never lay down. They can kill me, but I'll never shut up. This is my life." And by *this* she means her personal fight to stop the killing and the lawlessness that the coal industry brought to Martin and Mingo Counties.

This month, here on the Kentucky side of the Tug Fork, Carter has gotten some help from the state's new commissioner of the Department of Vehicle Enforcement, Greg Howard. Appointed by the newly elected Republican governor, Ernie Fletcher, Howard has instructed his department to start weighing and fining overweight coal operators.

This may have been a direct reaction, on Fletcher's part, to departed Democratic governor Paul Patton, whose family is heavily invested in strip mining. For Patton, there was simply no money in enforcing the law on haul weights. His family and friends would do much better hauling overweight loads of coal and then bidding on contracts to resurface the damaged roads. That way, they got you coming and going.

But whatever Fletcher's motives for enforcing the law, it quickly became clear to the drivers that not only were they going to pay the fines while the coal companies pled ignorance, they were also not going to make any money. With no union, the drivers have been at the mercy of the coal companies, who pit them against one another for the lowest haul fee. As the wife of one trucker said, "You can put two coal drivers in a room full of whores and they'll come out screwing each other." So today the result is about fifty rigs sitting idle along the side of I-80, across from the Leslie Resources coal tipple. The drivers have lifted their empty beds and set up lawn chairs along the roadside. They are demanding that Horizon Natural Resources, Leslie's parent company, double their haul rate per ton. And two days from now, their demands will be met.

For now, the police are waving traffic along, so I drive on to Lost Mountain to find out what a month of heavy blasting has done. It is a hot, nasty climb through the blackberry brambles that cling to what's left of the eastern flank. But little is left. Just above this scrub, a huge black cavity reaches down to the 10 seam. Behind it, a gray bench flattens off at the 11 seam, while ugly scree washes down the side of it. One chestnut oak leans perversely out from the bench, as if its momentary reprieve were a final reminder of who's in charge here.

With no one to haul the coal today, the black pit sits empty. My boots sink into the loosened soil that has been pushed out of the way. There are no more natural scents here, only the faint smell of chemicals. I follow some dozer tracks up to a twenty-foot-wide haul road that leads all the way around the backside of the mountain. Now the entire forest down below has been severed from the small clump of trees still hanging on at the summit. An eight-foot coal seam runs along the inside wall of the haul road. This is number 12, the highest seam, the last to go.

On the western side of the mountain, orange and yellow fuses wind through the black beds of coal like broken spiderwebs. Only a few dozers are working down below, grading the rubble of the valley fill. Haphazard mounds of black and gray rock are piled everywhere. Empty explosive boxes litter the site. I step over fissures in the ground where spoil has been piled back over empty pits and compacted.

So much spoil has been piled around the pits here on the west side of the mine that I can work my way around to the back of Lost Mountain without being seen by the dozer operators. But when I start to climb my usual path to the summit, I realize with a shock that the entire eastern ridge side, where last month I was so tangled in blackberry briars, is nearly gone. All the vegetation has been shaved away, and a dozer has cut a long scar all the way up to the summit. The oak-pine forest that once surrounded the mountaintop is now only a narrow strip of trees. What was once a gently sloping ridgetop is now a long vertical rockface, dropping hundreds of feet and jutting out over the gray shelves below.

For the first time, I approach the summit with a real sense of urgency. I may not see it again. Next month, these capstones may be gone, these chestnut oaks erased.

From the top, my eye follows the long gash that the dozer carved down the eastern side. Brown sandstone gives way to gray haul roads and black coal pits. Beyond the mine site, dozers have already started grading the lower region of the valley fill, compacting the rock that once held up this mountaintop.

I sit down on the only capstone that hasn't been dislodged from the top of Lost Mountain. Surrounded by several chestnut oaks, the stone is cold, and covered with lichen and Virginia creeper. The last remnant of the forest descends behind me, down the backside of the mountain. Two hundred feet down in front of me I can see nothing but this huge blister left on the land. The ugliness ends abruptly at the thin gray line that Highway 80 draws across the lower horizon. Beyond that perimeter, low green ridges flow away into the distance, an undulation of densely textured green waves. Clouds cast dark blue shadows that settle in their hollows. What was once the bottom of a huge sea is now the bottom of the sky.

These two landscapes, divided by the highway, illustrate two ways of thinking about the natural world that Wendell Berry recently set out in an essay called "Two Minds." Berry begins by making a distinction between a *rational* and a *sympathetic* mind. He admits that such a dichotomy risks oversimplification, but it is nevertheless a useful distinction. The rational mind is objective, analytical, empirical. It believes in industry, individualism, and an economy where profit is always the bottom line. The sympathetic mind is not unreasonable, but it favors the organic, the intuitive, the wild. "Its impulse is toward wholeness," Berry writes. Conversely, the rational mind is governed by the equation "knowledge = power = money = damage," and nowhere is that more obvious than where I am sitting right now. If the mountains on the other side of the highway truly represented the sympathetic mind, they would not necessarily be cordoned off as absolute wilderness. Hunters would still find their protein in those woods. And the trees would be logged in a manner that sustained the forest for generations to come. Herbs would be harvested there, and shiitake mushrooms would be raised on cut logs under the damp shade of hemlocks. The mountains would still be a resource, but they would also remain whole, an ecosystem left healthy and intact. To think with the sympathetic mind is to think, in Aldo Leopold's famous phrase, "like a mountain." When Leopold was working for the Forest Service in the 1910s, hunters and foresters killed every wolf they came across. The wolf was a predator, and a dead wolf was a good wolf. The result? Deer populations grew to the point that they nearly stripped Southwestern mountains of all their vegetation. Leopold took something important from that experience. We cannot think of the natural world as a collection of individual parts: wolves, trees, water, coal. We must think of the entire system as a whole; we must think like a mountain. "Only the mountain," wrote Leopold, "has lived long enough to listen objectively to the wolf."

WHICH SIDE ARE YOU ON? (PART 3)

"Bloody Harlan County" earned its epithet in the '30s when the powerful Harlan County Coal Operators Association hired Chicago gangsters and local felons to terrorize miners who were suspected of joining the United Mine Workers of America. The sheriff, his deputies, mayors, even then Kentucky governor Flem Sampson all did the coal companies' bloody bidding. The Harlan County Coal Operators Association paid the salaries of all the sheriff's 164 armed deputies, 64 of whom had been indicted and 27 convicted of felonies, including murder. As a result, Harlan County coal miners became the last workers in the United States to earn the right to collective bargaining.

The decade-long struggle began in 1931, when coal operators announced a 10 percent wage reduction for miners whose families were already living on the brink of starvation. Sensing an opening, the

United Mine Workers of America rallied together thousands of miners at a Pineville theater and urged them to organize. Coal company spies, however, had infiltrated the gathering. The next morning, all the miners who had attended the meeting were summarily fired and their families evicted from their coal camp homes. Many of them withdrew to the noncompany town of Evarts.

On May 5, as ten deputies were escorting a nonunion miner to a Black Mountain mine, rifle and shotgun fire rained down on them from the surrounding woods. The miner and three deputies were killed in the gauntlet. Two days later, Governor Sampson sent 370 National Guardsmen to restore order in Evarts. In the end, forty-three miners were charged with murder. The union drive collapsed when the governor refused to order the reemployment of union miners, and the Red Cross refused to offer them relief. Ultimately, as historian John W. Hevener wrote, "the miners were to be starved back into the pits."

Sheriff John Henry Blair's deputies, called "gun thugs" by the miners, continued their reign of intimidation as a new union, the Communist-backed National Miners Union (NMU), began setting up soup kitchens in Harlan County. In his fascinating memoir *Growing Up Hard in Harlan County*, G. C. Jones remembers seeing "men floating down Martin's Fork, bloated beyond recognition" and "deputies rolling bodies of striking miners over the edges of the road, watching them roll and slide into deep, dark wooded ravines."

Sam Reece, a local organizer affiliated with the NMU, nearly met such a fate one night in 1931, when Sheriff Blair sent a group of deputies to Reece's house. When they found only his wife, Florence, and their seven children home, the men ransacked the place, then waited in the woods to ambush Sam. Somehow, Florence Reece got word to her husband to stay away. During the long standoff, so the story goes, she tore a page from a wall calendar, and on the back of it wrote what would become perhaps the most famous union song in the world—"Which Side Are You On?" The third stanza goes

They say in Harlan County
There are no neutrals there.
You'll either be a union man
Or a thug for J. H. Blair.

The chorus repeats four times the troubling question "Which side are you on?" Then the song ends:

Don't scab for the bosses,
Don't listen to their lies.
Us poor folks haven't got a chance
Unless we organize.

There were a lot of poor folks. In a three-month period of 1931, the midwife Aunt Molly Jackson, who would later achieve fame as a union balladeer, saw thirty-seven children die in her arms from TB, pellagra, and other illnesses related to poverty. She earned the moniker Pistol Packin' Mama for the time she found seven children in a coal camp crying because their mother had nothing to feed them. Molly made for the company commissary, where she picked up a twenty-four-pound bag of flour and proceeded toward the door. When the storekeeper objected, Aunt Molly pulled a .38 from under her coat and said, "Martin, if you try to take this grub away from me, God knows that if they electrocute me for it tomorrow, I'll shoot you six times in a minute." That year she wrote the song, "I Am a Union Woman." On an Alan Lomax Library of Congress recording from 1961, you can hear Aunt Molly's strained voice singing a cappella:

The bosses ride fine horses
While we walk in the mud.
Their banner is the dollar sign
While ours is striped with blood.

And while singers like Aunt Molly Jackson and Florence Reece did focus some national attention on the plight of Harlan County miners, nothing seemed to change in Harlan County. In 1933, the National Industrial Recovery Act guaranteed employees the right to join a union. The Harlan County Coal Association countered by creating bogus company unions and again using gun thugs to discourage any other union activity. The NMU never got a foothold in Kentucky because of its affiliation with "reds" and "outside agitators," and the UMW seemed helpless in the only state where company-paid deputies ruled by what amounted to martial law.

Finally, on February 9, 1937, a car stopped in front of the home of union representative Marshall Musick. The driver fired into the family's living room, instantly killing Musick's teenage son. It was as if the nation had suddenly had it with Bloody Harlan County. Kentucky governor A. B. "Happy" Chandler, who had earlier supported Sheriff Blair, convened a special session of the state legislature to reform the private deputy system. In 1938, Kentucky became the last state to abolish the deputy mine-guard system. In addition, President Roosevelt formed the National Labor Relations Board, which began investigating employer violations of workers' civil liberties—specifically, the right to organize. A federal grand jury indicted twenty-two coal companies, the sheriff, and twenty-two operators for conspiring to deprive miners of the right to collective bargaining. Though volatile outbursts from the jury forced the judge to finally declare a mistrial, the coal operators had spent almost $1 million on their defense, and they didn't want a retrial. They were ready to sign union contracts. G. C. Jones remembered, "It was a great victory for the union. The thugs that weren't given prison terms left Harlan County. Some of the thugs that left soon met a torturing death by unknown persons."

Wages rose and the abuse of miners halted after 1938. But a declining demand for coal and the mechanization of the industry meant that mine employment in Harlan County dropped from 15,864 in 1941 to 2,242 in 1961. Families fled to Northern cities. But in 1974, the UMW did succeed in reorganizing 15,000 nonunion miners. In the Oscar-winning documentary *Harlan County U.S.A.*, Barbara Kopple chronicled that fight between the UMW and Duke Power over a contract that would guarantee health care benefits for underground miners. Duke Power took the position that the occupation of deep mining had nothing to do with black lung disease. The strike was long and contentious. Duke Power hired latter-day gun thugs to wave pistols around and generally intimidate men and women on the picket line.

At one point in the film, during a UMW rally, an old woman in a red-and-white gingham dress stood up in front of the crowd of striking miners. The place erupted with applause. Everyone knew it was Florence Reece. She told the crowd that her father died in the mines. She said her husband, Sam, was dying at home of black lung.

"I'm old and I can't sing very well," she offered by way of apology. "But you can ask the scabs and the gun thugs which side they're on." Then in a high, tinny voice, she started to sing:

Come all you good workers,
Good news to you I tell
Of how the good old union
Has come in here to dwell . . .

The strike ended in a contract, but it proved to be one of the last union successes in eastern Kentucky. Since Kopple's documentary was made, UMW membership has dropped to one-fifth of what it was then, and there's not one union mine in Harlan County today.

At ten in the morning late last August, I took up a position on the corner of Vine and Limestone Streets in Lexington, Kentucky, across from the U.S. Bankruptcy Court. That's where about a thousand angry miners were about to converge. I watched them approaching from several blocks away, wearing bright yellow United Mine Workers T-shirts. They brandished cowbells and megaphones, chanting, "Hey hey, ho ho, bankruptcy laws have got to go." By the time the first marchers reached the Community Trust Bank building, which housed the bankruptcy court, their line stretched out over ten blocks.

When the first wave of protesters had filled the sidewalk space in front of the bank building, the marchers pointed their megaphones up at the third floor, where Judge William Howard was about to finalize a deal that would allow the bankrupt Horizon Natural Resources, the parent company of Leslie Resources, to sell its properties to New York billionaire Wilbur Ross for $786 million. The snag was that Ross's

Newcoal LLC only wanted to buy Horizon if it first canceled all health care and retirement benefits for miners at six of its union mines. The crowd began to chant,

"Who has the power?"

"We have the power."

"What kind of power?"

"*Union* power."

It was an admirable show of strength, even if the facts belied the sentiment. Two weeks earlier, Judge William Howard, appointed to his post by George H. W. Bush, ruled that Horizon did not have to honor union contracts for 3,300 working and retired miners. Today, Newcoal and Horizon would finalize the deal and all of these miners would be left with nothing.

Someone was singing "Solidarity Forever" as I squeezed to the front of the crowd. Then, when a man in a suit tried to exit through the front glass doors, a group of miners, all in their fifties and sixties, abruptly took a seat on the sidewalk, blocking the exit. Confused, the man inside retreated. A few others miners joined the clutch of men on the ground. Another grabbed a megaphone and offered a prayer for the future of all union miners—and for the immediate prospects of these seventeen men, stoically seated with their arms crossed. By the time the prayer ended, the seated miners were surrounded by helmeted metro police officers. One cop motioned a miner to his feet. The man stood up and put his hands behind his back. The officer cuffed him and they started toward a police van. Everyone in the crowd cheered the miner.

When Cecil Roberts, the president of the UMW, was motioned to his feet, a huge cheer erupted and the crowd shouted, "Cecil! Cecil!"

He raised his fists in the air, then dropped them behind his back. It went on like that. One by one, the seventeen miners were cuffed and led away as the crowd cheered and yelled its support.

As the miners stood waiting to be loaded into the van, a local TV reporter asked if it was worth going to jail over this issue. They all said it was. I asked if they thought their actions would make any difference. "It's a start," said William "Bolts" Willis, president of Local Union 8843. "We have over a hundred thousand years of working experience at our coal mine, and this is what we got for it."

One of the policemen was taping the whole thing with a video camera, to prove, I suppose, that no one was being mistreated. And no one was. For all of the shouting, this was a scripted, civil affair. The men made their point and the police made their arrests. No one suffered any delusions that things would be made right on this day. It would all get reported on the six-o'clock news, and then the miners—many of whom suffer from black lung disease—would go home and start trying to figure out how they would pay for doctor visits and prescription drugs.

August 2004

LOST MOUNTAIN

August is the month of the nine-foot-tall wildflowers. Pink joe-pye weed and purple ironweed lean over the back roads of Perry County as I drive past. As usual, I park my truck on the side of the road beneath Lost Mountain. But something has changed: When bankrupt Horizon Resources sold off its assets last month, International Coal Group (ICG), owned by billionaire Wilbur Ross, bought the Lost Mountain permit. Leslie Resources has reorganized as a subsidiary of ICG and will continue to mine this coal without having to worry now about former creditors.

And something else has changed: I make it halfway up the eastern slope when I notice the beeping of the dozers sounds much closer than it has before. When I come around one of the old switchbacks, I see why. The entire eastern ridgeline has now been leveled all the way to the edge of this slope. A new valley fill is taking shape below. I work my way down the ridge side to take a look, and soon find myself slogging through an old dump site. There are bottles, coiled barbed wire, sheets of corrugated metal, exhaust manifolds. I step gingerly around the broken glass and black locust thorns. According to the coal industry, where I'm standing represents another good excuse for mining. The argument goes like this: These ignorant people have already trashed the mountains. They have dumped their garbage at the heads of hollows and run straight pipes from their toilets into the creeks. Thus, *it's already fucked up anyway.* Why not mine it? It's just a little more waste in the hollows, a little more acid in the streams. Such a conclusion, though it goes largely unspoken, is shared by many urban people when they think of the mountains of eastern Kentucky, and it suggests one of the reasons so little attention was paid to the slurry pond break in Martin County.

What it doesn't suggest—what it ignores—is the poverty that accounts for the lack of proper plumbing or proper waste disposal. Nor does it admit to the magnitude of the carnage created by strip mining. A man who junks his old water heater on one side of the mountain may still hunt and run his dogs on the other side. Granted, he may win no stewardship award, but he isn't destroying a watershed either. And unlike the executives of Massey Energy and Peabody Coal, he lives there. But beyond this question of scale, what the absentee coal companies also ignore is the rich tradition of families like Damon Morgan's, who farmed, hunted, and gathered herbs throughout these mountains. The literature of Appalachia *is* a literature of place. In no other work that I can think of does place—region—play so vital a role as in the novels and stories of Harriet Arnow, James Still, Robert Penn Warren, Denise Giardina, Lee Smith, and many other Appalachian writers. Yet the perception of this land and these people as damaged goods persists. As novelist Silas House has pointed out, every major mudslide in California received national media attention, yet the constant mudslides caused by valley fills in Appalachia are ignored. A nation of editorialists can get exercised about the environmental consequences of drilling for oil in the Arctic National Wildlife Refuge, but the much more destructive effects of mountaintop removal seem nowhere near their radar. Recently, Sonoma State University's annual Project Censorship report listed mountaintop removal as one of the top ten underreported issues of 2005.

When I finally get clear of the garbage and dense scrub, I can see a silt pond that has been dug at the bottom of the valley fill and lined with riprap. The pond is designed to catch sedimentary erosion before it washes down into the creeks below. It drains into a larger pond below the original valley fill. When I move closer, I notice that one bank of the lower pond is already scarred with the orange acid drainage that is trickling down from a culvert.

It's a hot hike back up the mountain. As I climb, I watch two dozers shovel topsoil off the ridge into the valley. When I reach the backside of the shelf where they are working, I crawl down through a tangle of uprooted trees that is now twenty feet thick. What strikes me is the suddenly fecund smell of all this topsoil that has all been shoved down over the ridge sides. I grab hold of exposed roots and pull myself up to a foothold where I can see and not be seen.

Several hundred feet away, what's left of the summit now stands isolated, like a butte rising suddenly in the Arizona desert. It is almost completely inaccessible, circled by a hundred-foot highwall on three sides. Back here on this eastern bench, I watch the dozers work for a while. It takes a thousand years to build twelve inches of topsoil on these steep slopes. But it will only take the dozer driver a few hours to scrape it all away. The "Methods of Operation" section of the Leslie Resources permit for Lost Mountain states that "existing topsoil which can be feasibly saved will be salvaged and stored." "Feasibly saved" is another one of those phrases, like "to the extent possible," that are so intentionally vague they mean nothing. Clearly, no topsoil is being saved on this bench, and by now there is plenty of space for it—about forty acres—at the flattened top of Lost Mountain. But in strip mining, expediency always carries the day. Coal is cheap because it is extracted with the least concern for the land that offers it up.

When SMCRA was passed in 1977, states were given the choice of enforcing it themselves or leaving it to the federal government. The coal industry fought for state regulation because, at least in this state, local legislators are more accessible and cheaper to influence. In addition, state control turns the federal minimum guidelines into the state's maximum standards. Consequently, as environmental lawyer Tom FitzGerald told me, "We do the minimum required under federal law." The coal industry sees no monetary value in saving topsoil, and so it sees no value at all.

This is just one example of how the true cost of coal is externalized onto the land and the people of Appalachia while absentee coal companies siphon off the profits. We also build deadly impoundment ponds rather than use a safer dry-filter-press system, which separates the water from the slurry, which is then buried; we plant grasses rather than native trees and call it reclamation; we bury streams under billions of tons rather than tunnel underground for the coal. And consequently coal remains, in the popular consciousness, "cheap energy." But the current price of coal tells nothing near the truth about the cost of air pollution, water pollution, forest fragmentation, species extinction, and the destruction of homes. Natural capital is destroyed and monetary capital is exported as quickly as the coal.

Walking back down the mountain, I listen to the cracking and falling of trees under the weight of the spoil as it is pushed down into the valley.

RFK IN EKY

On a cool September morning, I was standing in a small gravel parking lot outside the Vortex Community Church, a one-room white clapboard building that sits right off Route 15, northwest of Lost Mountain. In place of a steeple, a worn piece of plywood stood mounted on the roof, announcing, "Gospel singing every third Saturday night." About a hundred people were milling around the church. The men wore narrow ties, and jackets with thin lapels, the women silk hats and veils. One woman was decked out in a pink dress suit, pink hat, Jackie Kennedy sunglasses, and an incongruous pair of black leather boots. I had on an old shirt I had pilfered from my father-in-law's closet, along with a retro tie and a straw fedora. We were all supposed to look circa 1968, the year Robert Kennedy came to eastern Kentucky.

For one day, I was playing a very minor role in a very major production. *RFK in EKY* is a sprawling piece of performance art that had been four years in the planning. The script is a strange compilation of "found texts," namely, transcripts from Robert Kennedy's three-day visit to eastern Kentucky in February 1968. Admittedly, one doesn't think of eastern Kentucky when one thinks about experimental street theater. But John Malpede is not a conventional theater director. His own company in L.A. is made up of homeless people from Skid Row. Malpede conceived of *RFK in EKY* as a piece that would reenact, almost word for word, Kennedy's visit to the region. Malpede wanted to re-create the event in real time, at the actual places Kennedy visited. I had dressed accordingly and driven down with the vague idea that I might write an article for some theater magazine. I wasn't thinking that the arrival of the avant-garde in Appalachia had much of anything to do with mountaintop removal. It was only after the reenactment of Kennedy's speech to a group of college students at the end of that first day that I realized it had everything to do with it.

Kennedy had come to eastern Kentucky to conduct "field hearings on hunger," which were meant to assess Lyndon Johnson's War on Poverty four years after its implementation. Kennedy's first idea was to visit South Carolina, but Senator Ernest Hollings begged him not to, promising his own investigation into that state's poverty. So Kennedy came to the coalfields instead. (Harry Caudill once remarked ruefully that while the mountains of North Carolina had the Biltmore, and West Virginia had The Greenbrier, poverty was eastern Kentucky's most popular tourist attraction.)

Though Kennedy had not yet announced his candidacy for president, many, including Johnson, thought it was inevitable, and not a few considered the visit to eastern Kentucky mere grandstanding, political theater meant to show up LBJ. But people here don't remember it that way. Robert Kennedy was good-looking and charismatic, and seemed truly to care that many Appalachian children were starving. Rarely do coalfield residents speak of Johnson's first visit forty years ago, but almost everyone who was alive then has a story about when RFK came to town. He visited a one-room schoolhouse and a strip mine; he held Senate subcommittee hearings on employment and poverty; he visited a faltering African-American community in Hazard; he spoke at a small college high in the hills at Pippa Passes.

Over a four-year period, Malpede worked in cooperation with Appalshop, a grassroots arts organization from Whitesburg, to pull it all together. Founded in 1969, Appalshop has earned an international reputation for training local people to document, through film, radio, and theater, the cultural life of Appalachia. Its radio station, WMMT, the self-proclaimed "voice of the hillbilly nation," is fantastic. Malpede, for his part, started the Los Angeles Poverty Department—the LAPD—twenty years ago so that some of the country's poorest could dramatize their own stories and create their own public stages, even if those stages were shelters or rehab centers. Which is to say, both Malpede and Appalshop have plenty of street cred, and together, they were taking their brand of activist art to the streets.

In this case, the street was a typically narrow, unlined road where people were parking along the shoulder and moving with a bit of uncertainty toward the small white church. We all faced the same dilemma—we weren't sure when life left off and art began. My role was doubly confusing, since I was "playing" a journalist, but also doing my best to *be* one. The blurring of such boundaries, I realize, is what performance art is largely about—to get people to look at life in the same way they are used to looking at art, and thus to start talking about it. In this particular case, that discussion might concern the fact that the poverty rate in eastern Kentucky is exactly the same as when Kennedy came, jobs are almost as scarce, and the environmental destruction from strip mining is exponentially worse.

While we waited for RFK's arrival, I spotted Greg Howard, the director of Appalshop, and asked if he thought this was a celebration of progress the region had made since '68 or a dramatization that there hadn't been enough. "I think the point is to get people to ask just that question," Howard said. "Where are we now? What can we do about it? This performance seems like a better way to have that conversation rather than simply holding a panel or a local symposium."

Malpede cast only local people to reenact the parts, the lives of local people thirty-six years ago. And in a way, this kind of performance art is an extension of the strong oral culture that has always held sway in central Appalachia. "This is a place where you don't stand on ceremony too much and participation is the nature of the culture," Howard said. "You make your own fun. That's the point, right? You don't learn a musical instrument to just play it in your room." Nor do you construct a political drama to watch it played out within the confines of a theater or an opera house.

As we were talking, a '58 Edsel pulled up, orange with white fins. A man in a black suit and a black tie stepped from the backseat and started for the church. Handlers surrounded him, women came up to greet him. The show was on.

This Robert Kennedy, played by a community theater veteran, Jack Faust, didn't quite look the part. He was older and heavier ("I would make a good Teddy Kennedy," Faust had joked earlier). But he moved with an unmistakable bearing and everyone knew who it was.

When we had all crowded into the church pews, Frank Taylor, playing Congressman Carl D. Perkins, stepped up to a small podium. Wearing a brown suit and tie, he told this audience of his fictional, but very real, constituents that eastern Kentuckians were "the finest people in the world." He went on, "Of course we're poor, but we love our country and we love Senator Kennedy." We all clapped to that.

Faust stepped to the microphone as RFK and told the crowd, "We're here to find out whether the programs the government has begun are effective. We live in a country with a gross national product of eight hundred billion a year, and it grows by sixty billion a year. It's not acceptable, it's not satisfactory, that people do not have enough to eat." Everyone applauded. "And beyond that, in an area such as this, we must ensure that the young people stay home. That if they want to work, there is a job here for them. This part of the country has suffered, and it's unacceptable that it continues to suffer. If a person lives in poverty and wants to work, there should be a job, and if we can't do something about it, the government must do something about it.

"My brother knew that in times of crisis, the people of eastern Kentucky gave their blood, their energy, their courage to this country. And now we, the rest of the country, who live in so much affluence, have an obligation and a responsibility to ensure that the people of eastern Kentucky share in that great affluence."

Witnesses came forward and squeezed onto the small dais. One woman, Phyllis Buckner, was playing her mother Betty Terrill. She said she had six children and that her husband made $3 a day as a farmhand.

"How much?" Kennedy asked.

She repeated the figure.

"What is your total income for the month?"

"Sixty-nine dollars," she replied.

"For eight of you?"

"Yes."

"And what do your children have to eat?" Kennedy asked.

"Well, we have a right smart," Buckner said. "We raise our own hogs and have a cow and we always raise a garden."

Kennedy asked what she thought was the area's greatest problem. Buckner said she wasn't sure she could send her oldest two children to high school the following year because of the cost of books. Kennedy asked if there were jobs in the region. No, she said, the coal mines had shut down, and farming was only seasonal work. There was no industry, so all the young men went north if they wanted to work.

All the other witnesses had similar stories. They paid $50 for $70 worth of food stamps. They lost jobs to the mechanization of mining. They raised gardens and hogs and chickens and cows to keep food on the table. Nancy Cole's husband had "burned up in the mines." She was raising eight children on her own. Quietly, she recounted, "I got on welfare."

"Would you talk louder?" Kennedy asked.

"I was on welfare," Cole blurted, as if she had been forced to confess to an affair. But, she told Kennedy, "The people here are not sorry." They didn't want handouts. If there were jobs, they would work. But if there weren't jobs, and the government would not create jobs, Cole said, "Raise the welfare payments so people can live like human beings instead of living like . . ." Her voice trailed off.

When the hearing ended, I stepped into my own more improvisational role as a journalist. Outside the church, I stopped Don Doklage, a professional storyteller who had played John Akemon, a grassroots organizer, and asked what impact he thought *RFK in EKY* might have on the region. "I think these times, generally, parallel those times thirty-five years ago," he said. "We have the same economic problems, the same racism, the same involvement in world politics in questionable ways. So even though this event happened in '68, this reenactment has something to say to us today. I think any art, to be real art, has to do that."

We all got back in our cars, and the caravan drove twenty-five miles southeast before turning off onto a narrow two-lane that snaked fifteen miles up the mountains, toward an abandoned one-room schoolhouse that stands in the community of Barwick. Kennedy had wanted to meet some children. His aide, Peter Edelman, had organized the original itinerary in '68, but no one had anticipated the number of journalists from the national press that would descend on eastern Kentucky for those three days. Our own entourage mimicked that original press corps as forty cars and vans followed RFK along this winding stretch. And as was surely the case in '68, the families that live along this road have come out on their porches to ponder our convoy, perhaps having no idea that they are playing the part of earlier residents who must have shared the same incredulity.

The real Peter Edelman was part of our group as well, revisiting these spots for the first time since '68. In his recent book, *Searching for America's Heart,* Edelman reminisces about the years he worked with RFK, "a man who constantly sought out [injustice] and tried to right it." Edelman later served under Bill Clinton as an assistant secretary of

Health and Human Services, but resigned in 1996, when Clinton signed the "welfare reform" bill. Edelman thought that piece of legislation would punish millions of poor children whose only mistake was being born poor. "What President Clinton signed," he wrote, "was not a responsible policy but an abdication of responsibility." Today, nearly one in five American children still lives in poverty. In parts of eastern Kentucky, that number rises to one in two.

We all walked around mud puddles toward the one-room clapboard schoolhouse, the only building that remains from the original tour. Zona Akemon's family members, who still own the land where the school stands, had spent the last month restoring it to the way it looked on that February morning in 1968. There was still a potbellied stove in the middle of the room, and on each wooden desk there stood a photograph of one of the children Kennedy spoke with. Red paper Valentine's Day hearts had been taped to the windows, and portraits of George Washington and John F. Kennedy hung above the back door.

The room was dark and video footage of Kennedy talking with the children had been projected against the far wall. *RFK in EKY* has made no attempt to reenact what went on in this room on that day. For one thing, no one knows exactly what Kennedy told the children. It is clear from the news footage that they were terrified by all these strangers with cameras who had descended upon them. Perhaps sensing this, Kennedy simply knelt beside the desks and spoke quietly with each child.

As the video looped over and over, the school's teacher, Bonnie Jean Carroll, and its cook, Zona Akemon, stood before the now crowded room and talked about those times. Akemon and her husband John, whose stand-in I had spoken with earlier outside the church, had started the Grass Roots Program here in Barwick around the time of Kennedy's visit. They built a greenhouse, whose rusting frame stills stands on a hill behind the school. "We started plants so people could have gardens," Zona Akemon recalled. "We gave them jars and showed them how to can in a pressure canner." Before that, most people built fires outside under an old bathtub and canned that way, but the food didn't keep as long as with pressure canning. The Grass Roots Program raised money for a tractor and started a livestock program. "Each family got a cow," said Akemon, "and the first calf that their cow had, they had to give it back. That's how they paid for their cow. If they got a pig, they got to keep one from the first litter and give the rest back; that's how they paid for their pig." When director John Malpede asked Akemon why she and her husband had started the program, she replied, "We were just trying to help people to help themselves to do better. That's all."

What about the children, someone asked, what did they think of Kennedy? "They all thought he was real pretty," recalled their teacher, Bonnie Jean Carroll.

Meanwhile, art and life kept intermingling. As we filed out of the school, Zona Akemon approached Don Doklage and said, "I want to meet the man who says he's my husband." The impersonator smiled and stuck out his hand.

Walking back to the cars, I caught up with Peter Edelman, now a law professor at Georgetown University. "It's great to come back and relive it a little," he said. "It's good to see the people carry on. But it's sad to see people still struggling. We have over twice as much income as a country as we had back then, and it's all stuck at the top."

He acknowledged that the food stamp program is better now and that the elderly poor receive better treatment. "The elderly have political strength. They vote. So children are now the poorest age group."

Welfare, he said, is not the issue—poverty is the issue. He quoted some numbers: 35 million Americans live in poverty today, 3.4 million more than in 2000. There are over 15 million people whose incomes are less than $7,500 for a family of three. "The biggest reason people are poor is because they're working and they're not making enough money. That should be the easiest part." I mentioned a living wage. "It doesn't all have to come from the employer," Edelman said. "It would be a big jump to go to a living wage from what some people are making, so we need to have an earned income tax credit. But we don't even do very well with that, and if you're talking about people who are just really set apart from the rest of the country, it's so easy to forget about them."

I thought of Barbara Ehrenreich's book *Nickel and Dimed*, where she recounts her experiences impersonating a single woman with a limited education. She worked three separate low-wage jobs in three different parts of the country and found that she and her coworkers could hardly survive on what they made. Women she waitressed with lived in vans because they didn't have the down payment for an apartment; a cleaning woman worked with a broken ankle so she wouldn't be sent home without pay; Wal-Marts, such as the one recently built in Hazard on a former strip job, don't pay their workers extra for overtime; management, across the board, was cruel and petty. It struck me that Ehrenreich's project was itself a kind of performance art. She was, after all, only *acting* poor. But what she discovered was real enough: "The poor can see the affluent easily enough—on television, for example, or on the covers of magazines. But the affluent rarely see the poor. . . . The poor have disappeared from the culture at large, from its political rhetoric and intellectual endeavors as well as from its daily entertainment." On this day, the people of Barwick were enacting the fact that they exist as working, struggling Americans, and not as Jed Clampett's hillbilly kin or James Dickey's brutal rednecks.

Back in 1968, Zona Akemon had made lunch for Kennedy, and now, in 2004, several of us stopped to buy sack lunches from her grandchildren. I asked Edelman about the coal industry. "I remember back at one of the hearings in '68, Harry Caudill talked about all this money that is going outside the state, and those people own the politicians," he said. "They still do. Coal didn't exactly leave much behind in terms of helping the people who live here."

In '68, Kennedy drove east to a strip mine along Yellow Creek. The operators weren't exactly glad to see him. According to one man who was with Kennedy, "He had to get out and use a little gentle personal persuasion to even have a look at the mine site."

This time, the operators knew we were coming. In fact the cast of *RFK in EKY*, along with its vagabond audience, had been invited to the site by B & W Resources—with the stipulation that nothing negative be said about strip mining.

About forty cars climbed the muddy road up to the mine. We parked along the rim of a large pit where two front-end loaders were filling haul trucks with large blocks of sandstone. More than once during my visits to Lost Mountain I had thought of the strip mine as a perfect stage set for Theater of the Absurd. Those long gray benches looked so desolate and timeless—something out of Sartre's *No Exit* or Beckett's *Waiting for Godot*. But I doubt that was what mine supervisor Doug Melton was thinking when he invited this particular theater troupe up to his strip job.

He stood with one foot on the back bumper of his pickup while several of us asked him questions. There was no script for this stage of the performance. Whatever happened happened. RFK was leaning against his orange Edsel. And as a larger group of people formed around Melton, John Malpede quickly grabbed from his car an amplifier and a microphone. But when he stuck the microphone in front of Melton, Melton recoiled.

"No no no," he sputtered, "I don't want to be interviewed."

"It's just for amplification," Malpede explained. "So people can hear what you're saying."

And Melton had a lot to say. Holding the microphone at his chest, he said he was proud that B & W was a local company, proud of the fact that all 131 employees came from this or surrounding counties. All the money stayed in eastern Kentucky. "We're just a bunch of local people who work real hard to do what we do," he said. For him, that work had put two daughters through college and one through law school.

Every once in a while Melton had to stop talking as one of the huge haul trucks crept out of the pit and toward the hollow fill on the backside of the mine.

He said when oil prices go up, coal generally follows. Two years ago, it wouldn't have been worth it to extract the coal from this mountain, but given the volatile oil market, now it was. The plan was to mine from here to the highway, about a mile away.

The woman in the pink dress suit and the Jackie Kennedy sunglasses asked, "Given all of your experience, is there a better way of extracting coal?"

It was a pointed question, but Melton didn't flinch. "We're not interested in deep mining," he replied. "We're completely satisfied with the way we do things. Hopefully, ten years from now we'll be doing exactly the same thing."

He touted the benefits of level land, then added, "This will be a tremendous piece of property once we finish. When we walk out of here, it will be a showpiece."

When he was finished, Peter Edelman came over to where I was standing and said, "What did you think?"

I smiled and shrugged.

"He's good at it," Edelman said.

No question. I could have pointed to the valley fill on the other side of this ridge, which we weren't allowed to see. I could have mentioned forest fragmentation, species loss, flooding, the sham of reclamation. But that wasn't part of the performance. This was Doug Melton's stage, and what he had dramatized was his *humanness*. Here was a real guy, with a real family, just trying to make his way in the world. And he seemed *nice*. He had invited us up here, after all (in that light, the question I has asked earlier about whether or not this was a union mine—it wasn't—seemed downright rude). He wasn't a gun thug or a Horizon vice president, trying to take health benefits from miners. I could easily see myself talking with him over a beer about bluegrass music or University of Kentucky basketball.

Still, to me there seemed an incredible disconnect between what he was saying and where he was saying it—this scarified mountain ripped open with explosives. But for him, this was just where he came to work. And that, finally, is what made this scene an extraordinary piece of impromptu theater.

The Edsel led us all back down to the main road, then up a narrow two-lane that ends at the campus of Alice Lloyd College, where stately log and stone buildings stretch out along the banks of a headwater stream. This was Robert Kennedy's final stop on that first day in 1968. Like Kennedy, Alice Lloyd had come to Kentucky from Massachusetts. In 1916, she started a community center in Knott County, where a farmer had offered her his land if she would educate his children. In 1923, she started the college that now bears her name in hopes of educating young people, while urging them to stay in the mountains and work to improve their communities. All of this appealed to Robert Kennedy, of course, and he came in '68 to remind the student body that "service to community is part of the responsibility of a college education."

Thirty-six years later, around dusk, a newer auditorium had filled with a new set of students, this time in compulsory attendance for their history classes. But they all seemed willing enough to play along as Jack Faust took the stage in his black suit and tie. Reciting Kennedy's speech, he told the students, "You live in one of the richest places in the world. The natural resources here in eastern Kentucky are matched only by the riches you find in Africa in the diamond mines. The problem is absentee ownership. The great wealth of eastern Kentucky has been transferred out with little being returned and invested in eastern Kentucky." People applauded. "Investors," Kennedy went on, "have garnered great profit on the backs of the workers of eastern Kentucky. This is a situation that is intolerable. What we have to do is create an environment where the wealth generated by the natural resources of eastern Kentucky stays in eastern Kentucky." There was more applause.

RFK began urging the students to use their education not only for "self-betterment" but also for "community betterment." "Without community support, there is very little reason to attain an education," he said. "Because what do you have if you don't have a community worth living in?"

Then he opened the floor for questions. In 1968, the president of Alice Lloyd College had told the students not to ask Kennedy any questions about Vietnam, that he was tired of having to answer them. But true to that night in '68, a young woman immediately stood up and said, "Senator Kennedy, what about the Vietnam War?"

What followed was an eerie set of remarks that dissolved the lines not only between life and art but also between past and present.

"What we have," Kennedy said of the South Vietnamese, "is an ally in name only. We support a government that has no supporters. What we have is a government that without the support of the United States, would fall tomorrow."

Suddenly a young man jumped to his feet in the back of the room and shouted, "Do like Truman did—let's bomb them! Let's bomb the whole place, wipe it out."

A smiling RFK agreed that was indeed one option. Was there anyone else in the room who shared these sentiments? he asked. No hands went up. Was there anyone who thought the war should be escalated?

No hands. "How many people think we should begin our withdrawal from Vietnam?" Kennedy asked. All hands, save one, rose into the air.

"The policies of escalation in Vietnam have not consolidated our strengthened international resolve to resist aggression," he began. "What it's done is weakened other people's faith in our wisdom and our purpose." Spontaneous applause erupted. One began to sense that the students were not only watching a production, but beginning to consider Kennedy's words as relevant counsel in 2004.

RFK asked if the students had seen the recent photograph of a South Vietnamese police chief summarily shooting a Vietcong soldier in the head (the photo won Eddie Adams—who would die this month—the Pulitzer Prize in '68). "That picture," said Kennedy, "appeared on the front pages of newspapers throughout the world, and some of our oldest allies said more in sorrow than in anger, 'What has happened to America?' The reason we wage a war against Communism is not to become more like them, but to preserve the difference between us." There was more energetic applause.

"We were told that Vietnam would settle all of the difficulties in Asia," RFK went on, "that it would somehow preserve the security of the United States. But that is a wish based on mindless hope allowing us to justify the great sacrifices we have already made. The truth is, Vietnam will not protect America from its enemies.

"What we have is a situation where people at the highest level of government will not tell us the truth. They would rather not tell us the truth than risk the repercussions of what the truth may bring."

There was no question: This "found text," this old speech recontextualized as a work of art, had weirdly risen to the level of prophecy.

Finally, RFK brought the war home. "We cannot establish in Asia a great society if we are unwilling or unable to create one here in America," he said. The war had divided the nation and drained resources from places like eastern Kentucky. Today, 39 cents of every Kentucky tax dollar goes to the military. Four cents goes into education, and less than half a cent is spent on job training.

"Today, for the first time, I saw a strip mine," he concluded, "and the land was devastated. So what happens is, the people of eastern Kentucky are three-way losers. The minerals are gone, the money is gone, and the land has been despoiled."

When the performance was over, John Malpede called Peter Edelman to the stage to offer some historical context to Kennedy's speech. But before he started to speak, Edelman said, "I want to recognize that Ann Caudill, Mrs. Harry Caudill, is here."

Ann Caudill stood up to much applause. She said that Kennedy's visit had given her husband hope that a century of corporate oppression and raw poverty might finally come to an end for eastern Kentucky. And she remembered this: "When Robert Kennedy was about to get into the car to drive away, he turned around and he came back to my husband. He took him by the hand and he said, 'Mr. Caudill, we're going to come back and we're going to do something about all this.' And much has been done. But there is an awful lot more to do."

Four months after Robert Kennedy shook Harry Caudill's hand, he won the Democratic primary in California. Four hours later he was shot dead.

September 2004

LOST MOUNTAIN

It was one year ago this month that I first came to Lost Mountain. When I look back at the pictures I took then, I see dense stands of trees and rolling ridgetops painted orange and yellow by autumn coolness. Now I see a long gray plateau piled with mounds of wasted rock and soil. It's drizzling as I start up the eastern slope. Today is a Sunday, like a year ago, and the rain has probably kept even the smaller, weekend crews away. I don't see or hear any dozers or loaders. At about fourteen hundred feet, I begin walking along the top edge of a long highwall that marks the eastern boundary of the land permitted for mining. This cliff line drops about one hundred feet down to the number 10 coal seam, where several pyramids of coal stand ready to be loaded away.

I'm walking along a thin strip of soil here at the edge of the highwall that divides the strip mine from the forest. The oaks and maples descend down into the watershed on my right, and the highwall drops away abruptly to my left. The sharp contrast between these two landscapes, heightened by the fall color and the gray mine site, gives me the strange sensation that I am standing on the edge of Creation, on a thin membrane between the world and the not-world. It's as if the Creator had been busily composing this variegated forest and then suddenly knocked off for the day, right where I'm standing. Everything past this point is an abyss, a lifeless canvas, a preternatural void.

At the end of the highwall, I climb down onto the mine site. The wet coal crunches softly under my boots. I walk toward the former mountaintop, where I had parked my truck a year ago. I stay close to the northern edge of the mine, just in case one of the company's omnipresent white pickups should appear and I need to duck quickly down into the forest. Because all this earth has been churned over many times, my boots sink deep into the orangish mud. It's slow going. As a shortcut, I drop down into the woods, cross a narrow ravine, then climb back up through the inevitable blackberry brambles and young sassafras trees. At the top of this ridge, another long bench levels out below the northeastern side of the mountain. I slog across this coal seam, where coyote tracks appear inside the wider imprint of dozers. A muddy embankment rises at the back of this bench. There is no way to get any kind of a foothold, and my mud-caked boots now weigh twice what they did. I finally find a tree root sticking out a few feet above me and make a clumsy leap for it. Slowly I pull myself up onto a ledge where fallen trees lie scattered. I kick my boots against a stump and gingerly wade up through another thicket of briars.

Stepping over a final berm of spoil, I find myself standing where the capstones once sat. Now all the vegetation has been shaved away. A long yellow fuse winds up to what once was the mountaintop but is now only an awful black knob. I follow the fuse to the edge of that small plateau, leveled off at the number 11 coal seam. The wasted summit is now a series of tall gray mounds of rock piled to my right. They almost look like glaciers, shooting unnaturally up from this man-made desert, rising above small black pools of rainwater mixed with coal. To my left, the entire eastern ridgeline has been carved up and hollowed out; now it is only one wide black crater. And down in front of me, a gray bench has been turned to concrete by the heavy trucks that, over and over, have backed to its edge, then methodically dumped this mountaintop down its side. I'm standing in the middle of a wasteland, a dead zone, a man-made desert.

It won't always look this bad up here. Eventually, this spoil will be pushed down the ridge side, then sprayed with a mixture of grass seed and fertilizer. With any luck, the grass will take hold and keep the spoil from washing down the hillside. This fractured landscape at the top will be slowly leveled by graders, then it too will be seeded with the exotic lespedeza. According to the reclamation section of the Leslie permit, what was a mixed mesophytic forest will be turned into a "pasture." When I showed that part of the permit to a manager for the Office of Surface Mining who is known for his frankness, he shook his head and said, "It's a joke the way reclamation is done now, to have pasturelands just sitting on the top of a mountain."

But whatever this landscape becomes, the mountain is gone for good. Its trees are gone, its topsoil is gone, and its forest-dwelling species, many nearing dangerously low numbers, are gone.

From here, I take in the entire panorama of this blasted ridgeline, this eviscerated forest. I think for a moment that I might write a short poem, a eulogy to Lost Mountain. Nothing comes to mind. The ancient Chinese poets wrote out of a deep identification with their own mountains, one so strong that many of those poets are now remembered not by their own names but by the names of the mountain they ranged across. Here there is little left to identify with, nothing that seems the proper subject of poetry.

Not that I came to Lost Mountain for inspiration. Though I have been inspired by its songbirds, its watersheds, its wildflowers, I knew its fate a year ago, when I started wandering these flanks that no longer exist. I came to Lost Mountain for a firsthand education. I climbed to its summit again and again to see what can't be observed from below—the systematic destruction of an entire biological community.

In short, I came to Lost Mountain looking for what Aldo Leopold called "an ecological education." "One of the penalties of an ecological education," wrote Leopold, "is that one lives alone in a world of wounds." Certainly coalfield residents and their allies can feel quite alone when they face the enormous power, and sometimes the violence, of the coal industry. No one who inflicts wounds, from soldier to strip miner, wishes to be reminded of the fact. But to watch a 300-million-year-old mountain destroyed over the course of one year, and to view from a plane the number of mountains that have suffered the same fate, is to understand that this land is a badly wounded organism.

That the land *is* one organism, Leopold thought to be "the outstanding discovery of the 20th century." James Lovelock and Lynn Margulis reaffirmed this discovery in the early '70s, when their research into the temperature, composition, and oxidation rate of the atmosphere found that the earth does indeed act like one self-regulating macroorganism. Leopold chose another metaphor, one from the folklore of his native Wisconsin: the Round River. It was Paul Bunyan and his blue ox who discovered this mythic river that flowed into itself in a continuous cycle, and Leopold rediscovered it in the '50s with the advent of ecology. "Wisconsin not only had a round river," he wrote, "Wisconsin *is* one. The current is the stream of energy that flows out of the soil into plants, thence into animals, thence back into the soil in a never ending circuit of life." Margulis has even suggested an analogy between the circulation of blood through our arteries and organs, and the circulation of the earth's waters from clouds to streams to oceans to evaporation and back to clouds—another Round River.

The upshot of all this is that if the land and its atmospheric membrane behave like a single organism, then everything within that organism is serving some function—all the parts are working interdependently for the health of the larger organism. Even the wolves that prey on the deer are working to preserve the health of that larger entity, the mountain itself. For Leopold, whose influence on wildlife conservation and wilderness protection in the United States can hardly be overestimated, the health of that land depended on two things: stability and diversity. The stability of an integrated forest leads to the accumulation of soil fertility; soil fertility leads to biological diversity; biological diversity leads to "one humming community of co-operators and competitions, one biota." Of any biome, of any watershed, Leopold said we must ask two questions: "Does it maintain fertility?" and "Does it maintain a diverse fauna and flora?" These questions formed the basis of what Leopold called a "land ethic," which if successful, would "change the role of *Homo sapiens* from conqueror of the land-community to plain member and citizen of it."

Standing on this sterile ledge, I am surrounded by the work of conquerors, not members of any land community. No one who felt a responsibility to other citizens within a community would destroy its water, homes, wildlife, and woodlands. The difference between conquerors and community is the difference between the words *economy* and *ecology*. Both come from the same Greek root, *oikos*, meaning "house." But only ecology has remained such a study. A true case of home economics would, as Leopold said, make sure the place called home maintains its health and stability. To create an environment where mudslides, flooding, and slurry spills are common will not ensure a community's health. To bulldoze and burn a renewable resource—trees—will not ensure its stability. To tear a nonrenewable resource from the ground to provide short-term economic gain for the few and long-term environmental destruction for the many is undemocratic, unsustainable, and stupid.

We are, unfortunately, a nation that values technology and wealth much more than we value community, and the result is the wasted land that lies all around me. The twentieth century was a Faustian gamble that combined industrialism and greed to make us cash-rich and resource-poor. If our species is to make it through this century, the forces of science and technology must be tempered by two other forces: ethics and aesthetics. As Leopold observed, all philosophies of ethics, from Aristotle on down, are actually based on this ecological principle: "that the individual is a member of a community of interdependent parts." And as the cave art at Lascaux makes brilliantly clear, we are a species that has evolved to find beauty in the natural world. This trait serves—or should serve—an evolutionary purpose: we love what we find beautiful, and we do not destroy that which we love. What a strip job demonstrates, then, is the absence of any ethic or aesthetic.

It is more than a moral failure; it is a failure of the imagination—a failure to understand energy and employment alternatives that would preserve the integrity and the beauty of the Appalachian Mountains.

The fighting between conservationists and the coal industry—between an ethic and the economy—will rage on for years. That's clear. And the fight might have to get quite ugly before substantial, sustainable change occurs. We are not a country given to velvet revolutions. We will have to choose sides, it seems, to reach the point where we realize *there are no sides,* and that there are no sides because there is no *outside.* This small planet is all we have, and to continue on our current course will be to ensure that we all become outsiders. It is, I think, for this reason that former Czech president Václav Havel said we must "reconstitute the natural world as the true domain of politics." Ideology and arbitrary borders mean little when roofs won't stay on houses anywhere and people die of bronchial infections everywhere. It is time we stopped thinking like those who conquer mountains and started thinking like the mountain itself.

PART TWO

BEFORE THE LAW

After that first year, I stopped making my monthly visits to Lost Mountain. There wasn't much more to see or say, only more of the same. And since I no longer needed to move around Lost Mountain incognito, I decided to blow my cover. In December 2004, I filed a citizen's complaint with the federal Office of Surface Mining, alleging that the blasting on Lost Mountain was damaging the groundwater at the head of Lost Creek. It was just a hunch. But since filing such a complaint is one of the few recourses citizens have when fighting the coal industry, I wanted to exercise my right to such a visit.

Within hours of filing the complaint on the OSM website, my answering machine was full of messages from operators and regulators who wanted to discuss my allegation. People who a year ago wouldn't return my calls were suddenly happy to be of assistance. Brian Patton, a major player with International Coal Group (ICG) and the cousin of our former governor, offered to give me a tour of the site personally. When I mentioned that I thought the state regulators might give me a more objective understanding of the site, Patton accused me of "boxing him in." Filing a complaint, he assured me, "was no way to do business." I might have replied that I wasn't trying to do any business, but it was already clear that our conversation was going nowhere. When I asked how he obtained my unlisted number, Patton informed me that he "had his sources."

The December morning of the inspection was cold, gray, and rainy. Coffee in hand, I stood under the awning of the BP station at the bottom of Lost Mountain, waiting for the state regulators to arrive. Don Gibson, a representative from ICG, was there as well. We shook hands grimly and I tried to explain that I had nothing personal against his company—that I chose to write about this mine site because I just liked the name Lost Mountain.

"Lucky us," he mumbled.

When three men from the Department of Natural Resources arrived, I climbed into their four-wheel-drive Jeep and we headed for the site. When I asked if the University of Kentucky Wildcats had won the night before, the lead inspector, Jeff Taylor, asked, "Are you a UK basketball fan?" I said I was. "Well, at least we've got that in common," he said almost mournfully.

So this was the dynamic: No one wanted me here, and I could think of about fifty places I would have rather been. The operators obviously didn't want to see any writers. And the regulators didn't want to be shown up for violations they hadn't cited. There was a general sense of touchiness all around. I almost wanted to apologize to the regulators for putting them through all this: I wanted to assure them that I wasn't a bad guy, and wasn't out to make them look bad. This was the one day when I actually held a degree of power and leverage against a leviathan industry, but I found that I wasn't much enjoying it.

At the guard shack, the safety inspector asked if we were sober. He said alcohol and drugs were "getting to be a big problem" on mine sites. When we assured him that our cups held only coffee, we drove up the muddy haul road. The mine site looked worse than ever. So much gray spoil had been piled around the pits and highwalls that I couldn't see more than a hundred feet in front of me. Several times we took a wrong turn.

Finally, we stopped at a blast hole where men where loading ANFO into a sixty-eight-foot-deep cylinder. By now several more white trucks had arrived, and about ten more men from the Department of Natural Resources piled out to watch me watch the operators. Our boots sank deep into the mud, and the drills roared around us. Almost shouting, I asked the blasting operator how many pounds of explosives would be detonated at this hole. "I don't know," he said. "I'd have to figure it up." Back in the Jeep, one of the regulators assured me that he would figure it up before the blast went off.

We drove on to where I had alleged that blasting might be damaging groundwater. I pointed down into the black pit whose bottom couldn't have stood more than fifteen feet above the head of Lost Creek. I tried to lay out my case over the noise and the rain. One inspector said he thought my allegations held merit, another assured me that I didn't have a case. Either way, given the machinery and the highwall terrain, it would be impossible to take a water sample from here; we would have to approach the creek from below.

Was there anything else I wanted to see up on the mine site, Taylor asked me. I said I'd like to take a look at the hollow fills. In two of the mine inspection reports, the stability of the hollow fills is questioned. One reads, "Care must be taken that material dumped is 70% durable"—that is, rock. Another reads, "Monitor hollowfill closely to ensure that organic material is not placed in it." I had seen organic material—trees—being buried under the spoil dumped by haul trucks. And I had watched tons of nondurable material—topsoil—pushed down into these fills. When I asked Don Gibson, who had been trailing me closely, where the topsoil was being saved—because the permit map showed that it would be—he said Leslie had received an Alternate Topsoil Variance from the state agency and could dump all the topsoil it wanted. An officer at OSM later told me that such variances are handed out like coupons by the state agency. But even so, that didn't solve the problem of the hollow fills' durability. I turned to the mine inspector, and asked if these fills looked to him to be in compliance. He paused and said, "They're close."

I wasn't sure what "close" meant, but an officer at OSM had already told me, in a moment of candor, that there wasn't a legal hollow fill in Kentucky. Later on, I called Jack Spadaro, who had, after all, headed the Mine Health and Safety Academy, and he confirmed that assessment. "None of the fills are 70 percent durable rock," he said. "And all are in violation of the Clean Water Act because they are filling in stream channels with mine wastes. But officials at these agencies continue to issue the illegal permits."

Why were none of the companies being cited for violations? Because no one inspects the hollow fills after they have taken shape. According to Clyde Cook of Kentucky's Department of Natural Resources, core samples of the strata at the proposed mine site are taken before any permits are issued. Then a computer model is used to simulate the makeup and geometry of the proposed fill. That's it. Once the permit is approved, none of the regulatory agencies demands further core samples to show that the valley fills actually measure up to the 70 percent durability test. Consequently, the coal companies have license to dump potentially dangerous material without a real threat of legal reprisals.

According to Spadaro, the valley fills of central Appalachia are the largest earth structures in North America. No one can really predict the future of a hollow fill because no one can predict the weather. A saturated hollow fill could quickly become a mudslide. And as environmental lawyer Tom FitzGerald told me, the fills could give way far into the future, long after the five-year bonding period, meaning that the coal company responsible for the slide could not be held financially liable.

Spadaro believes that massive mudslides and rockslides washing off valley fills are potentially as big a danger as impoundment pond breaks. All across eastern Kentucky and southern West Virginia, mudslides have swept through homes and destroyed the property of people who cannot get flood insurance. In addition, the enormous amount of erosion caused by the fills has clogged streams and rivers, greatly reducing their storage capacity. The result is seven "one hundred year" floods in the last three years. These floods roared down the mountainsides and killed fourteen people. Back in 1981, a four-foot wall of rock and coal slurry rolled down a mountain in Harlan County. It pinned Nellie Woolum to the wall of her kitchen and she was asphyxiated. In

August 2004, a boulder rolled down a hillside in Inman, Virginia, and crushed to death a sleeping three-year-old, Jeremy Davidson. Matt Mining, the company responsible, was fined $15,000.

It appealed the fine.

Before I left Lost Mountain on that gray December day, I led the guys from Natural Resources up over the deadfalls and slick cobble of Lost Creek. It was slow, wet going, and one hydrologist pointed out sensibly that with all of the rainfall, we probably wouldn't get a very good gauge of the groundwater itself. Still, Taylor seemed determined to make sure all my requests were met. So he and I slogged up the creek to the large beech that stood at its source. By this point, water was pouring down from above the source as well. Nonetheless, we duly gathered four bottles of running water and carried them back downstream. The four of us trudged back out of the woods and up to the Jeep, where Don Gibson was also waiting in his truck. His window was down, so I walked over and attempted a joke. "Hey, bad news," I said. "We just found an endangered species of fern."

"Aw, it don't matter," he replied. "We're almost finished here."

CONCLUSION

The year 2004 ended badly for those of us who saw George W. Bush's reelection as a critical missed opportunity to redirect this country's energy and environmental policies. More to the point, it ended badly for the mountains—for the streams and the people of Appalachia. Now that the Bush administration has changed the wording in the Clean Water Act so that mining waste has become benign "fill," and now that the price of coal has doubled over the last year because of volatile oil markers, the permitting process for more mountaintop-removal permits will certainly quicken. There will be more dangerous floods and more mercury in the water, more damaged wells and more dangerous coal trucks, more carbon in the air and fewer trees to sequester it.

In an essay I mentioned earlier, "The Last Americans," Jared Diamond asked why the Mayan kings of the eighth century didn't recognize that they were hastening their own demise by destroying their natural resources and producing uncontrollable levels of waste. The answer is that it wasn't in their short-term interest to do so. Furthermore, Diamond wrote, "it's difficult for us to acknowledge the wisdom of policies that clash with strongly held values." In the United States, the value of individualism, largely understood as one's unfettered right to use resources, clashes with the values of conservation and preservation of those resources for future Americans.

Three days after Bush's victory, Vladimir Putin signed the global, CO_2-reducing Kyoto Protocol, thereby ratifying the treaty that could go into effect only if 55 percent of countries producing greenhouse gases agreed to its mandate. And while the United States produces twice as much carbon dioxide as Russia, Bush made it clear in the presidential debates that he would not sign the Kyoto Protocol because it could "cost American jobs and stifle economic growth." In other words, short-term decision making will continue to rule the day, though the long-term effects of those decisions could be disastrous.

As the 2005 energy bill makes clear, we as a country will continue to rely heavily on oil, and given the volatility of that market, we will rely increasingly on coal and nuclear power. Serious measures to move toward alternative energies were shuttled by the bill. This is obviously bad news for Appalachia. And it is bad news for everyone downstream. There is a certain insanity (I choose the word carefully) about perpetuating a global economy based on limitless growth when that growth is powered by finite resources—in the case of the energy bill, fossil fuels. Economist Herman Daly has pointed out that economics textbooks rarely mention words such as "environment," "pollution," "depletion." It is as if the neoclassical "growth" economists live on

Laputa, the island in *Gulliver's Travels* that floats untethered above the continents, filled with mathematicians who can think only in abstract terms with no correlation to the land below. Furthermore, the mainstream economic model refuses to calculate and internalize the damage it does to the natural world, thereby further depleting and degrading resources. As Daly writes, "The economy's *throughputs*—the flow of raw material inputs, followed by their conversion into commodities, and finally into waste outputs—[must] be within the regenerative and absorptive capacities of the ecosystem." But a strip mine is the clearest example one could find of how our current economy fails miserably at accounting for the true cost of "growth."

I, however, do not intend to end an unfortunately bleak book on a note of despair. So consider what the Nobel Committee decided around the same time Americans were deciding in favor of George W. Bush. The Norwegians awarded the 2004 Nobel Peace Prize to a Kenyan woman, Wangari Maathai—for planting trees.

Twenty years ago, Maathai's countryside, like the mountains of central Appalachia today, was on the verge of desertification. Poor rural women had to walk farther and farther to collect less and less wood for heating and cooking. So Maathai mobilized a legion of those women to start planting woodlots, known as green belts, throughout their villages. Soon there was more shade, less erosion, cleaner water, cleaner air, ample firewood for cooking, and jobs. Maathai's Green Belt Movement started solving problems of poverty, malnutrition, pollution, and women's rights. Maathai brilliantly used the example of a forest community to reestablish human communities across Kenya.

The Nobel Committee's recognition of Maathai's work, and its social ramifications for improving the lives of poor communities, is an important signal that Appalachia—indeed, all American communities—should take very seriously. There is so much abandoned, unreclaimed mine land in the United States that the Office of Surface Mining doesn't even try to account for it in terms of acreage. And a visionary organization, the Appalachian Regional Reforestation Initiative (ARRI), now has developed a successfully tested plan to bring forests back to strip mines. Based on experiments conducted by Clark Ashby in the '60s, foresters have found that the best way to reintroduce trees on a strip-mined mountain is simply to leave the soil uncompacted by bulldozers. If haul trucks dumped four-foot-deep rows of spoil across a mine site and planted tree seedlings in those mounds, even without *any* topsoil, the rate of growth would double that of seedlings in their native forests. Why? The loose soil gives their roots plenty of room to sink their tentacles, so the trees grow faster. Within five to eight years, their crowns would start to close, according to OSM field manager Patrick Angel. This man-made forest would mature faster than a natural forest because it cuts out about fifty years of early successional growth. That is to say, one can start planting late-succession trees like oaks, walnuts, and other valuable hardwoods right away. Those can be accompanied by faster growing pines, which could be harvested sooner for moldings or pulpwood. In a short amount of time, a sustainable local industry could take root on these barren plateaus. And if locally owned secondary industries, like furniture and cabinetmaking, took hold, the region's economy might begin freeing itself from absentee coal moguls. Trees, as I've mentioned, sequester much more carbon dioxide from the air than grasses. And pieces of furniture constitute what are known as "carbon sinks"—that is to say, the carbon in a piece of furniture, unlike that in paper or firewood, is going to stay there. As a final recommendation for this form of reclamation, the loose soil piled on the mine sites would collect much more rainwater than compacted valley fills, and flooding problems would ease. While such a plan does nothing to stop mountaintop removal, it at least returns the forest's intelligence to a landscape that has for too long been wasted by human arrogance.

In the past three years, in Kentucky alone, fifteen hundred acres have been reclaimed—and to my mind, truly reclaimed—using this method of reforestation. Over a million trees have been planted. But there are obstacles. Landowners and regulators alike have grown accustomed to the rolling grassland monocultures that constitute most post-mining land use. Initially, piles of spoil dumped on a site and dotted with seedlings doesn't look like reclamation at all. So the operators and even the regulators are afraid of it; and for the first few years, the landowner's view (if indeed he or she can see it) is not exactly sublime. In addition, as Patrick Angel told me, "Coal operators are very suspicious. If you try to propose something new, something innovative, it's hard to get operators to sign on." It just looks to them like more regulation.

But ARRI has calculated that the costs of reforestation—in terms of eliminating time spent running a dozer over and over spoil to compact it and purchasing nitrogen-rich fertilizer—would actually be cheaper. And nothing in SMCRA prohibits this kind of reclamation. One simply has to convince dozer operators that they do not need to turn spoil into concrete, and convince coal operators that this is a chance to actually do the right thing. Furthermore, if our policymakers had the vision to create what might be called a New Deal for Appalachia, using money from the Abandoned Mine Land Fund to train a generation of local men and women to reforest and manage abandoned strip mines, then the chronic problems of poverty and addiction would disappear from the mountains.

Still, a legion of public-spirited tree planters will not change a mindset that favors individual acquisitiveness over communal responsibility. In the end, Appalachia is a region that suffers, like the country as a whole, from a larger spiritual crisis. I think back to Mickey McCoy's yard sign:

GOD WAS WRONG
SUPPORT MOUNTAINTOP REMOVAL

As a consumer-driven culture, we have chosen to no longer think of the world as God-given. It's too inconvenient. Instead, here in Kentucky, our forests and streams are supposed to be protected by a Department of Natural Resources, because that's all we see them as—simply a resource. We too seldom see *value* in the natural world, whether aesthetic or intrinsic; we only see something we can use, even if that means using it up. We no longer see ourselves as part of a greater whole, a world so vast and mysterious that it deserves our reverence alongside our scientific probing. In America today, the "environment" is almost wholly other. We are over here, and it is over there. We are in the air-conditioned mall; it is hot and crawling with bugs. And anyone who prefers the out-there is an "environmentalist," that oddly dressed guy who thinks this diminutive planet might be worth saving.

The twentieth-century theologian Martin Buber made a crucial and influential distinction between two kinds of relationships: one based on an I–It principle, the other on an I–You way of thinking. The I–It mindset is reflected in the situation I have been describing: there is a subject and an object and each is alien to, separate from, the other. In the I–You experience, something different, something stranger happens. One loses a sense of the well-defined ego, the self; as a result, the other, the You, begins to seem not so alien anymore. Indeed, in some mysterious way, the You starts to feel like more of the I, and both the I and the You seem caught up in some larger interconnectedness. The thirteenth-century Buddhist master Dogen said, "To study Buddhism is to study the self. To study the self is to forget the self. To forget the self is to be enlightened by all things." Mystics call this experience "satori," or enlightenment; modern scientists simply call it ecology. In a forest, trees are photosynthesizing carbon into oxygen every spring. Whether I feel a spiritual connection to that forest or not, the fact is that it is keeping me alive. It is part of me whether I want to admit it or not.

The fact that the United States emits 36 percent of the world's carbon but will not sign the Kyoto Protocol on carbon emissions shows just how many of us don't want to admit just how entrenched we are in an I–It experience of the natural world. One of Wendell Berry's recent books is titled *Life Is a Miracle*. Why are we as Americans so stubbornly immune to understanding the world as something miraculous, as something imbued with spirit, as something worth preserving?

A thorough answer would require another book devoted to that question alone. But I want to sketch a quick outline of several historical developments that set the country on this track. At the heart of this dilemma is a radical dualism that severed spirit from matter, and matter from mind.

The modern world began with, and because of, the Age of Enlightenment. René Descartes, to his own satisfaction, proved humankind's existence by positing a thinking mind separate from all other matter. This divide provided a context for the mechanistic universe expounded by Isaac Newton: There were fixed, intractable laws that moved matter. The rational trumped the miraculous. The universe wasn't a sacred manifestation, as primitive creation stories led us to believe; it was more like a clock. And furthermore, while Descartes was separating mind from body, John Locke was separating the individual from society. The individual, he argued, had basic, inalienable rights, and these preceded that individual's responsibility to the collective.

Because of the rapid advances of the Enlightenment, the dominant path of Western societies over the last few hundred years has been toward the accumulation of scientific knowledge and consequently toward fields of specialization. The result, according to the philosopher Alfred North Whitehead, is that "we are left with no expansion of wisdom and with great need of it." Wisdom, for Whitehead, comes out of a broader, balanced development, which leads to an understanding that the parts are interrelated aspects of a larger whole. At nearly the same time, in the 1930s, Whitehead and Aldo Leopold were stressing that our culture's increasing emphasis on individuation and specialization was causing us to lose sight of the fact that the natural world is one organism in vital need of all of its symbiotic, complementary parts. To understand this and to act accordingly would be, for Whitehead, an important step in the direction of wisdom.

No one will fight to save something one does not love; Leopold identified that human trait long ago. Our spiritual crisis is that we, as individuals, too often cannot see beyond our own inflated narcissism to love something whose value cannot be immediately translated into monetary terms or human uses. While I have suggested that such thinkers as Locke, Descartes, and Newton led us to this point, I'm not suggesting that scientific values are harmful in and of themselves. What is disturbing and destructive is that our ethical values, our spiritual values, and our aesthetic values have not kept pace with science. Whitehead, a mathematician as well a philosopher, felt this strongly. And he thought this imbalance might be resolved through . . . poetry, particularly the poetry of William Wordsworth. "His consistent theme," wrote Whitehead, "is that the important facts of nature elude the scientific method. It is important therefore to ask, what Wordsworth found in nature that failed to receive expression in science. I ask this question in the interest of science itself." The answer was, in a word, *value*. A scientist, said Whitehead, can explain the exchange of atmospheric vapors that give color to a sunset, but only a poet like Wordsworth can convey the aesthetic, the spiritual, the inexplicable value of that same sunset. Or mountain. Or stream. Or bobcat. And, Whitehead went on, because the poet sees values beyond those of the merely applied sciences, the poet understands the natural world not as a machine but as an organism with intrinsic values and laws more mysterious than we might have otherwise accounted for.

In our time, the poet who has most fervently taken up the cause of

Whitehead's poetic values in the fight against the "technology = progress" paradigm is Wendell Berry. Consider how the following poem, "Manifesto: The Mad Farmer Liberation Front," brings Whitehead's philosophy into a contemporary context. The poem begins with the mad farmer addressing the modern American:

Love the quick profit, the annual raise,
vacation with pay. Want more
of everything ready-made. Be afraid
to know your neighbors and to die.
And you will have a window in your head.
Not even your future will be a mystery
any more. Your mind will be punched in a card
and shut away in a little drawer.
When they want you to buy something
they will call you. When they want you
to die for profit they will let you know.

Consider the logic here. We shut ourselves up in large, air-conditioned suburban houses (38 percent larger, and with fewer occupants, than in 1975), and as a result, often do not know our neighbors. Because we so often choose to live away from work, shopping, and entertainment, we must drive nearly everywhere we go. And because it takes a lot of gas to keep all of those cars and SUVs moving, we send young Americans to Iraq to "die for profit." It seems so insane to the farmer, he has gone mad. Or rather, his refusal to participate in such a system made him appear mad to all who simply accept the status quo. Yet that final reality that young men and women would have to die for something as trivial as profit shakes the farmer out of his despairing cynicism, and his tone abruptly shifts:

> *So, friends, every day do something*
> *that won't compute. Love the Lord.*
> *Love the world. Work for nothing.*
> *Take all that you have and be poor.*
> *Love someone who does not deserve it.*
> *Denounce the government and embrace*
> *the flag. Hope to live in that free*
> *republic for which it stands.*
> *Give your approval to all you cannot*
> *understand. Praise ignorance, for what man*
> *has not encountered he has not destroyed.*
> *Ask the questions that have no answers.*
> *Invest in the millennium. Plant sequoias.*
> *Say that your main crop is the forest*
> *that you did not plant,*
> *that you will not live to harvest.*
> *Say that the leaves are harvested*
> *when they have rotted into the mold.*
> *Call that profit. Prophesy such returns.*
> *Put your faith in the two inches of humus*
> *that will build under the trees*
> *every thousand years.*

To do things that "won't compute" is to adopt, as Whitehead might have said, a system of values outside those of the technologically driven consumer culture. One thing that doesn't compute within that logic is to "invest in the millennium"—to think beyond short-term economic interests, to plant trees, as the Menominee Indians do, that someone *else* will harvest. The mad farmer instructs the reader to reject the linear, industrial model that begins with finite resources and ends with waste, but instead to embrace the forest's cyclical system where rotted leaves turn to natural fertilizer. "Call that profit," says the farmer. "Prophesy such returns." The profit/prophet homophone in this line, along with the pun on "returns," suggests that real profit is that which naturally returns, as leaves do in a forest. Last, the mad farmer urges, put faith in topsoil, not in the bulldozers that would shovel it off a mountaintop.

Listen to carrion—put your ear
close, and hear the faint chattering
of the songs that are to come.
Expect the end of the world. Laugh.
Laughter is immeasurable. Be joyful
though you have considered all the facts.
So long as women do not go cheap
for power, please women more than men.
Ask yourself: Will this satisfy
a woman satisfied to bear a child?
Will this disturb the sleep
of a woman near to giving birth?

I have a friend, Jenny Williams, who lives in Hazard, Kentucky, and who derives immense satisfaction from being a mother to her seven-year-old son, Carson. Unfortunately, Jenny lives with her husband, Scott, and Carson beneath a hollow fill that has recently been cut and cleared by Leslie Resources. The forested headwater hollow where Jenny used to take Carson hiking is now a waste, a gray terrain of bulldozer tracks, spoil, and uprooted trees. Carson no longer sleeps in his bedroom, which sits at the back of the house, closest to the hollow fill; he sleeps with Jenny and Scott. But still Jenny has to go to bed every night thinking about the possibility of a flash flood that could bring mudslides and rockslides crashing down the hill toward her family. It disturbs her sleep. The mad farmer's manifesto concludes:

Go with your love to the fields.
Lie easy in the shade. Rest your head
in her lap. Swear allegiance
to what is nighest your thoughts.
As soon as the generals and the politicos
can predict the motions of your mind,
lose it. Leave it as a sign
to mark a false trail, the way
you didn't go. Be like the fox
who makes more tracks than necessary,
some in the wrong direction.
Practice resurrection.

Science without compassion, science without ethics, has given us the modern war machine, the industrial farm, the dead zone in the Gulf of Mexico, the strip mine. What this science has left out is everything the mad farmer stands for, and it is crystallized in the stunning last line—"Practice resurrection." What does it mean? Many things. Among them: practice and emulate the seasonal resurrection of the forest, understand and enact the miraculousness of *this* world, practice waking up to a world that is itself a miracle, plant a tree.

One seventeenth-century philosopher who did not fall in step with Locke and Descartes was Baruch Spinoza, a Dutch Jew who was excommunicated from his synagogue in 1656 and then set out to reconcile all religious and philosophical dualisms that formed the basis of Descartes's thought. The first divide Spinoza bridged was that between God and the natural world. He reasoned that because God was infinite, there could be no substance outside of God. God was self-created and self-creating. And because nothing existed outside of Him, and all being was of a single substance, then *God was Nature*. Thus, when Spinoza refers to God in his great work *Ethics*, he uses the phrase *Deus sive Natura*—God-or-Nature. Not only does God exist, God exists *right here*.

One can see how Spinoza was about four hundred years ahead of the Gaia hypothesis, which says the earth functions as a self-regulating, dynamic organism. As Stuart Hampshire puts it, "To say that God is the immanent cause of all things is another way of saying that everything must be explained as belonging to a single and all-inclusive system which is Nature." Spinoza simply made no distinction between the Creator and the Creation. God is the unique creator (*Natura Naturans*) and the unique creation (*Natura Naturata*). Some contemporary scientists have given the name *autopoiesis* to this self-organizing characteristic in nature. The simplest example of it would be the living cell.

CONCLUSION

Two billion years ago, free-floating bacteria combined to form what Lynn Margulis (whose theory this is) calls "bacterial confederacies." These confederacies gradually formed a thin membrane to hold them together. Then the bacteria turned into the oxygen-using mitochondria, and a command center, the nucleus, took shape. The eukaryotic cell was born, or rather, self-made. The cell took up residence inside larger organisms with their own protective membrane. These flora and fauna took up residence within forests and other ecosystems that also acted as even larger self-regulating organisms. And all of these biomes were protected by an even larger fortress, the atmosphere. The poetry of this is stunning when one considers how the macrocosmic planet and the microcosmic cell reflect such similar patterns of self-creation. As the rather dour evolutionist Richard Dawkins wrote, "Not only is Dr. Margulis's theory of origins—the cell as an enclosed garden of bacteria—incomparably more inspiring, exciting and uplifting than the story of the Garden of Eden, it has the additional advantage of being almost certainly true."

A similar poetic mirroring occurred when the human eye first viewed the Earth from outer space and saw something that looked very similar to its own cornea. Margulis has likened this experience to the image of Narcissus first looking into the river. Only this time, he sees not only the image of himself. He also sees the saw grass, bream, and pickerelweed. He sees every cell of his being reflected in the image of a small blue planet. He looks at the Earth and sees that it is one being, one self-regulating organism that, like the human body, needs all its symbiotic parts. He might even be moved to quote Charles Darwin: "We cannot fathom the marvelous complexity of an organic being. . . . Each living creature must be looked at as a microcosm—a little universe, formed of a host of self-propagating organisms, inconceivably minute and as numerous as the stars in heaven." To understand the universe as a divine macrocosm, and the earth, the human body, the nucleated cell as both reflections, or microcosms, of that *Deus sive Natura* and integral components within that Divine Organism is indeed an inspiring and ethically challenging creation story. It places us back in this world, where we no longer pray to escape this vale of tears but rather, to borrow a word coined by Gerard Manley Hopkins, to *inscape* more intensely into the experience of the natural world. The human mind is no longer separate from matter, as Descartes had it. Rather, as Spinoza wrote, "The greatest good is the union that the mind has with the whole of nature."

It is not my aim here to convert readers to pantheism (though some might find the recent theological reworking of the term—*panentheism*—less objectionable, since it sounds less like nature worship and more like the worship of God *in* nature). Whether one believes that the Creator made the Creation or that the Creator is the Creation, both beliefs require an act of reverence for the natural world (and as for atheists or agnostics, most of the ones I know care far more about the environment than the religious people I grew up around). But there is something fundamentally modern about Spinoza's philosophy: we see it in the basic principles of ecology, we see it in Margulis's science, and we find it in E. O. Wilson's biophilia hypothesis. The latter says this: *Human beings have an innate emotional affiliation with other living organisms.* Wilson points to the fact that more people visit zoos in a year than attend all pro sporting events combined. Moreover, human beings spent 99 percent of our evolutionary history as hunter-gatherers, depending on the natural world for our survival. It would be very strange if we hadn't treated large game animals as sacred and painted brilliant pictures of them in the caves at Lascaux. "In short," wrote Wilson, "the brain evolved in a biocentric world, not a machine-regulated one."

That we now live in such a highly mechanized world, and that such an artificial environment is the source of much depression and violence among *Homo sapiens*, is a theory that stretches from Alexis de Tocqueville to Thoreau to Paul Shepard to the Unabomber. It has been well documented, most recently by psychiatrist Peter C. Whybrow, that while Americans are four times more affluent than during the '60s, we have shown no measurable gains in happiness. In fact the opposite is true: We are more depressed, more medicated, more frazzled than at any other time in our short history. The more we are anesthetized by material wealth, the farther we stray from our biophilic selves. We move from house to garage to car to work to mall to gym to house again with little regard to our ancestral homeland—ancient savannas at the edge of vast forests.

Henry Salt was one of the first thinkers to extend the realm of ethics beyond the realm of the human. In his 1935 book, *The Creed of Kinship*, Salt declared, "The basis of any real morality must be the sense of kinship between all living beings." Seventy years later, I've tried to show in this book why Salt was right, why this very simple maxim should be the principle that guides us through the twenty-first century. While that sense of kinship among all living things can be explained scientifically through molecular biology, it will only be a force for change, a *moral force*, if it is understood by the *individual*. No one wants to be told what to do: turn off lights, drive less, recycle. But if a desire to change the way one consumes limited resources comes out of an inner conviction, a deep feeling of conscience, then it is not too late for a real transformation of our culture.

Frank Lloyd Wright, the American architect who seriously tried to collaborate with natural settings, once wrote, "The actual difference between 'individual*ism*' and individuality of a true democracy lies in the difference between selfishness and noble selfhood." Individualism is the right to passively consume in pursuit of a happiness based on convenience; individuality is a more creative attempt to invent oneself and one's life based not on commercial influences but on a more direct, intuitive combination of knowledge, passion, and responsibility. One is based merely on self-interest; the other is based on conscience. To me, that is really the crux of our American dilemma. We must choose, first as individuals and then as a collective, to reject selfish individualism for the individuality of *noble selfhood,* however each of us understands that term.

Last year, at the Lexington counterrallies about mountaintop removal, one man standing in front of the Kentucky Coal Association headquarters held a sign that read "Every American Born Will Need 561,477 lbs. of Coal in a Lifetime," as if that number was simply a mathematical constant, as if there were no alternative. But such an assumption is complaisant, unimaginative, and ultimately self-destructive. If one were to imagine as an inner portrait Leonardo's famous drawing of man with his outstretched arms and legs, one might conceive of a fourfold human being made up of these components: empirical knowledge, spiritual knowledge, ethical knowledge, and a capacity for aesthetic appreciation. This fourfold consciousness would, I believe, mark the beginning of a critical turn from selfishness toward noble selfhood, from a narrow understanding of progress to a more meaningful definition based not only on technological innovation and material gain, but also on the other three components of this inner self that we so often ignore. Material gain, speed, and convenience are the most dominant forces within this country, and they have done much to crush the spiritual, ethical, and aesthetic elements of our nature. If we understood the natural world as a spiritual presence, we would also see that all living things are kin to us. If this realization led to a moral attitude toward the natural world, then our destructive behavior would change. We would change. We would become more fully human. And we would recognize the natural world not merely as a resource, but as something much more profound—what Thoreau liked to call the Poem of Creation.

Acknowledgments

This book grew out of an experiment in experiential learning—namely a course my colleague Randall Roorda and I created at the University of Kentucky, something we call the Summer Environmental Writing Program. Every summer, we take a dozen or so students from all over the country to Robinson Forest and spend a month hiking, botanizing, canoeing, talking, and always writing. Robinson Forest is home to one of the last working fire towers in the state, and from the top of that tower one sees that the state's most pristine watershed is almost entirely surrounded by the scars of mountaintop removal. Standing at the top of the tower a few years ago, I decided I couldn't write about the beauty of Robinson Forest without also writing about the industrial economy that threatens to destroy it.

It was never my intention to become a student of mountaintop-removal mining, or to spend two years of my life wandering strip mines. Wendell Berry recently wrote, "I am a man mostly ignorant of the things that are most important to me." I share that sentiment, and I am sure people from the coal industry will be quick to point out places in this book where they perceive ignorance. That's fine. I will be happy to correct any technical errors I may be guilty of. What I do know, and what is beyond dispute, is that the mountains of central Appalachia are disappearing fast as a result of that industry's irresponsibility and greed.

Because I was not trained as a biologist or a mining engineer, I owe debts of gratitude to many who helped educate me about the mountains of my home state. Scientists often get a bad rap from us in the humanities, and I must say that, to judge from the men and women I have met throughout my fieldwork, that reputation is undeserved. John Cox, Jim Krupa, Dave Maehr, Phil Crowley, Deborah Hill, and many others have been extremely generous in helping me understand the subtleties of the mixed mesophytic forest, along with the forces that are working to destroy it.

When I first pitched an idea about mountaintop removal to John Jeremiah Sullivan, a senior editor at *Harper's* magazine, he said I needed a better narrative. "What if you followed one particular mountain from beginning to end?" he suggested. "You could call it something like *Death of a Mountain*." And out of that casual suggestion came a *Harper's* folio of that name, and then this book. John left *Harper's* to write full-time on the day I signed a contract for that article. But Roger Hodge very ably edited the 40,000 words I sent him down to 12,000. I am also grateful for the generous spirit and fact-checking expertise of Claire Gutierrez.

I am indebted to Jin Auh at the Wylie Agency for taking an unknown writer and guiding me through the mysterious world of New York publishing.

It was Sean McDonald at Riverhead who wrestled *Lost Mountain* into this final form, and who pushed me to confront blind spots within the text. That is to say, he found gaps and biases I could no longer see after a hundred readings, and he gently nudged me toward a better book. For that I am grateful.

As for the indefatigably generous Wyatt Mason, who read several drafts of *Lost Mountain*, I can say only that any writer would be extremely lucky to befriend a reader and a critic of such enormous goodwill.

As for matters more of the heart than the head, I have dedicated this book to two of my best friends, Scott Lucero and Jenny Williams, along with their son, Carson, for plying me with great food and ample drink, and giving me a bed, on the many nights that I spent in eastern Kentucky researching this book. Somehow, Jenny, Scott, and I escaped graduate school relatively unscathed, and their friendship as much as their culinary skills have been a great sustenance to me.

Because of the coal industry's reputation for intimidation in central Appalachia, I'm afraid that I caused my wife, Mary, far too much anxiety while I was away working on this book, and for that I am sorry. But Mary believed in me as a writer long before anyone else, and she knew this was a book I had to write. She edits everything I write, she knows the mistakes I will make before I make them, and she generally keeps me from embarrassing myself in front of other editors. My gratitude for that and a thousand other things is profound and compound.

Finally, my friend and mentor Guy Davenport died last January. I wish he had lived to see the publication of *Lost Mountain,* because he spent so much time teaching me, through his own writing and through conversations, how to shape the English language, how to elevate rhetoric to the level of art. He was, as far as I'm concerned, the greatest prose stylist of his generation. Every one of his sentences is a lesson, an event, a thing to be admired and pondered. It was simply through geographical luck that I was given the opportunity to study with him at the University of Kentucky, but over the last twenty years we became good friends, and I miss him terribly.

Notes

As is probably obvious, much of the information for this book was gathered through firsthand observation and interviews. Other sources are listed below:

Introduction

p. 3 **line at the top:** www.osmre.gov.

p. 4 **(EPA) estimates that at least seven hundred miles:** The Draft Environmental Impact Statement (EIS) on Mountaintop Removal Mining and Valley Fills of 2004 was prepared by the U.S. Army Corps of Engineers; the U.S. Environmental Protection Agency; the U.S. Department of Interior, Office of Surface Mining; U.S. Fish and Wildlife Services; and the West Virginia Department of Environmental Protection.

The New Canary

p. 7 **cerulean warbler populations across Appalachia are plunging:** Ted Williams, "Mountain Madness," *Audubon*, May–June 2001.

September 2003—Lost Mountain

p. 12 **"These Lost Creek miners":** Quoted in Shelly Romalis, *Pistol Packin' Mama: Aunt Molly Jackson and the Politics of Folksong* (Urbana and Chicago: University of Illinois Press, 1999), p. 27.

October 2003—Lost Mountain

p. 23 **Colonies of liverworts:** My thanks to ecologist Phil Crowley for educating me on the sex life of liverworts.

p. 23 **Bill Caylor:** "Stop Portraying Coal Industry as Bogeyman," *Lexington Herald-Leader,* June 30, 2003, p. A8.

p. 24 **"The outstanding scientific discovery":** Aldo Leopold, "The Round River," *The Sand County Almanac* (New York: Ballantine, 1970), p. 190.

p. 25 **30,000 deaths per year:** Clean Air Task Force, *Death, Disease and Dirty Power,* October 2000, pp. 3, 5, cited in Barbara Freese, *Coal: A Human History* (Cambridge, MA: Perseus Books Group, 2003).

p. 25 **asthma has risen:** Andy Mead, "Children's Health Tied to Environment," *Lexington Herald-Leader,* October 14, 2004, pp. C4f.

p. 25 **climatologists found record-high levels:** According to Russell Schnell, deputy director of the National Oceanic and Atmospheric Administration, in a 2004 study cited in Charles J. Hanley, "Another Step Toward Global Warming?" *Bergen County Record* (New Jersey), March 21, 2004.

p. 25 **forests worldwide have shrunk:** Lester R. Brown, *Eco-Economy* (New York: W. W. Norton, 2001), p. 55.

p. 25 **12 percent of the world's birds:** Ibid., p. 11.

p. 26 **A forest, by contrast, can store:** Ibid., p. 29.

Which Side Are You On? (Part 1)

p. 27 **In 1998, a group of West Virginia environmentalists:** The case was *Bragg v. Robertson,* Civ. No. 2:98-0636, according to the EIS.

p. 27 **A year later . . . coal and conservation:** Ken Ward, Jr., documented this series of events in "Mountaintop Removal Damage Proved: Bush Administration Proposes No Concrete Limits to Mining Permits," *West Virginia Gazette,* May 30, 2003.

November 2003—Lost Mountain

p. 34 **my fellow Kentuckian:** Wendell Berry, "Conserving Forest Communities," *Another Turn of the Crank* (Washington, DC: Counterpoint, 1995), pp. 41–44.

p. 35 **a formidable botanist named Lucy Braun:** Charles E. Little, *The Dying of the Trees* (New York: Penguin, 1995), pp. 145–64. Ronald L. Stuckley, "Women Botanists of Ohio," paper delivered at Annual Meeting of the Council on Botanical and Horticultural Libraries, May 20–23, 1992. Carolyn V. Platt, "Sisters in the Science Wing," *Timeline,* May–June 2002.

The Power to Move Mountains

p. 38 **According to the Kentucky Department of Natural Resources:** "Postmining Land Use," http://www.surfacemining.ky.gov/educateinfo/miningky/post/.

"Was It All by Design?"

p. 46 **The EPA excavated five thousand tons:** http://cfpub.epa.gov/supercpad/cursites/ccontinfo.cfm?id=0405125.

p. 47 **an elaborate shell game:** Ted Williams reported on the Dean brothers' activities in "Strip-Mine Shell Game," *Audubon,* November–December 1992, pp. 48–55. The "bad" standing of the Deans' bankrupt companies can also be tracked through the Office of the Kentucky Secretary of State: http://www.kysos.com/corporate2.

p. 48 **There, in *National Mining Association v. Department of Interior:*** "U.S. Appeals Court Voids Ownership and Control Regulations," *Engineering & Mining Journal,* July 1999, p. 32.

NOTES

January 2004—Lost Mountain

p. 53 **In the spring of '63:** The following series of events is recounted at the beginning of Harry Caudill, *The Watches of the Night* (Boston: Little, Brown, 1976), pp. 3–57.

p. 54 **"This was the only time":** Ibid., p. 15.

p. 54 **Appalachia's poverty rate:** Appalachian Regional Commission 2004 report. www.arc.gov.

p. 55 **Scientists calculate:** Jonathan Fowler, "Humans Draining Earth of Life," Associated Press, October 22, 2004.

p. 55 **In his essay "The Last Americans":** Jared Diamond, "The Last Americans," *Harper's,* June 2003, pp. 43–51.

p. 56 **Wilson warns:** E. O. Wilson, *The Future of Life* (New York: Vintage, 2002), p. 29.

p. 57 **a wood-products plant that received over $100 million:** Jason Bailey and Liz Natter, *Kentucky's Low Road to Economic Development* (Lexington, KY: Democracy Resource Center, 2000), p. 25.

p. 57 **this industrial park:** Ibid.

p. 57 **The state levies:** "Kentucky's Coal Severance Tax: Looking Back, Looking Forward," *Kentucky Source,* September 2003, pp. 14–23.

p. 57 **a computer call center:** "Economic Loser," *Lexington Herald-Leader,* July 3, 2003, p. A12. "Computer Support Company to Close Hazard Call Center," *Lexington Herald-Leader,* June 27, 2003, p. D2.

p. 58 **"Distance negates responsibility":** Guy Davenport, "The Master Builder," *The Hunter Gracchus* (Washington, DC: Counterpoint, 1996), p. 152.

p. 58 **Pakistani children:** Sydney Schanberg, "Six Cents an Hour," *Life,* July 1996.

p. 58 **Immigrant workers hack off fingers:** Eric Schlosser, *Fast Food Nation* (New York: Harper Perennial, 2002), p. 174.

February 2004—Lost Mountain

p. 71 **On August 8, 1967:** "Sunday Explosion Rips Strip Mine Operation," *Hazard Herald,* August 8, 1967, p. A1.

p. 72 **The violence peaked:** Caudill, *My Land Is Dying,* p. 87.

p. 74 **"I was born one mornin'":** George Davis of Hazard, Kentucky, wrote "Sixteen Tons"; he resented that Merle Travis changed some of the chords and lyrics. I quote from Travis's version of the song.

p. 75 **"The interior landscape":** Barry Lopez, "Landscape and Narrative," *Crossing Open Ground* (New York: Vintage, 1989), p. 71.

On Bad Creek

p. 81 **a native Kentucky schoolteacher:** Richard B. Drake, *A History of Appalachia* (Lexington: University Press of Kentucky, 2001), p. 145.

p. 82 **the Tennessee Valley Authority:** Caudill, *My Land Is Dying,* p. 75.

p. 82 **Mrs. Bige Ritchie:** Ben A. Franklin, "Hill People Join to Ask State to End Coal Operation," *New York Times,* July 1, 1965, pp. A1f.

p. 82 **the woman who became known as the Widow Combs:** Caudill, *My Land Is Dying,* p. 80.

p. 82 **an eighty-year-old coffin maker:** Ibid., pp. 75–76.

p. 83 **"I got seven shells in this gun":** Anne Lewis, *To Save the Land and People.* Appalshop Films, 1999.

March 2004—Lost Mountain

p. 88 **eight off-site flyrock violations:** "Office of Surface Mining 2004 Oversight Report for the State of Kentucky," Department of Interior, Office of Surface Mining (Washington, DC: GPO, 2004).

p. 88 **Oat Marshall:** Roger Alford, Associated Press, April 15, 2003.

p. 91 **70 percent more electricity:** Penny Loeb, "Sheer Madness," *U.S. News & World Report,* August 11, 1997, p. 30.

What Is a Flying Squirrel Worth?

p. 96 **The Cumberland Plateau:** Richard H. Yahner, *Eastern Deciduous Forest* (Minneapolis: University of Minnesota Press, 1995), p. 114.

p. 97 **By contrast, in a study:** J. F. Taulman and K. G. Smith, "Home Range and Habitat Selection of Southern Flying Squirrels in Fragmented Forests," *Mammalian Biology* 69 (2004): 11–27.

p. 99 **According to the World Conservation Union Red List:** Brown, *Eco-Economy*, pp. 69–72.

p. 99 **"If the decision were taken today":** Wilson, *The Future of Life*, pp. 101–2.

April 2004—Lost Mountain

p. 103 **Franz Kafka's short story "Before the Law":** Franz Kafka, *The Penal Colony*, trans. Willa and Edwin Muir (New York: Schocken, 1961), pp. 148–49.

p. 104 **During each year of his term:** See two very enlightening pieces of investigative reporting by Phillip Babich: "Shafted," December 11, 2003, and "Dirty Business," November 13, 2003, both in *Salon*.

p. 104 **Since the 1977 Surface Mining Control and Reclamation Act:** *Federal Register* 69, no. 4 (January 7, 2004), Proposed Rules.

p. 106 **Here is a characteristic poem:** Wang Wei, "Bird and Waterfall Music," trans. Kenneth Rexroth, *The New Directions Anthology of Classical Chinese Poetry* (New York: New Directions, 2003), p. 69.

p. 107 **So it was a minor epiphany:** George Constantz, *Hollows, Peepers, and Highlanders: An Appalachian Mountain Ecology* (Missoula, MT: Mountain Press, 1994), pp. 42–44.

p. 108 **Biologist Greg J. Pond:** "Effects of Surface Mining and Residential Land Use on Headwater Stream Biotic Integrity in the Eastern Kentucky Coalfield Region," study conducted for the Kentucky Department for Environmental Protection and the Division of Water, July 2004.

p. 110 **This coniferous-looking fern:** Barbara Freese, *Coal: A Human History* (Cambridge, MA: Perseus, 2003), pp. 17–18.

p. 111 **But one has only to go:** I am grateful to hydrologist Chris Barton for providing me with these numbers.

Acts of God

p. 112 **In 1912, the railroad:** Harry Caudill, *Night Comes to the Cumberlands* (Boston: Little, Brown, 1963), p. 93.

p. 112 **"Though he might revert on occasion":** Ibid., p. 99.

p. 114 **That part about treating your neighbor right:** Roger Alford, Associated Press, December 11, 2002.

May 2004—Lost Mountain

p. 125 **the price of coal per ton:** Lee Mueller, "A Very Good Time to Be Selling Coal," *Lexington Herald-Leader,* January 23, 2004, pp. C1f.

p. 125 **"In the modern world":** Rachel Carson, *Silent Spring* (New York: Crest, 1964), p. 17.

Whitewash in Martin County

p. 128 **In October 2000:** Lee Mueller, "Coal Slurry Pours into Two Streams in Martin County," *Lexington Herald-Leader,* October 12, 2000, p. B1.

p. 128 **For months following the Martin County disaster:** The *New York Times* finally ran a front-page story on Christmas Day, 2004.

p. 131 **An independent group of researchers:** Their report can be read at: http://www.anthropology.eku.edu/MCSPIRIT/PDF/Final_Report.pdf.

p. 132 **Coldwater Creek certainly deserved:** Lois Gibbs, *Mother Jones,* September–October 2004, p. 53.

p. 134 **After the Martin County spill:** I interviewed Spadaro to gather much of the following information, but the aforementioned article, "Dirty Business" by Phillip

Babich, was essential to my research, as was Clara Bingham's "Under Mined," from the January–February 2005 issue of *Washington Monthly.*

p. 135 **The investigators discovered:** Babich, "Dirty Business."

p. 135 **Then Scott Ballard:** "A Look at Testimony in Slurry Spill Investigation," *Lexington Herald-Leader,* April 28, 2004.

p. 136 **"The industry has always been good to me. I just hope that I've given back as much as I've received."** Oklahoma City *Journal Record,* June 19, 2003.

p. 137 **The rap sheet on Lauriski:** I am grateful to a host of researchers at the Democratic National Convention who, via FIOA, put together a timeline of Lauriski's meetings with coal executives and lobbyists. Their findings were gathered in the unpublished "The Loyal Servant: Lauriski Pushes Mining Industry Agenda."

pp. 138–139 **When the bogus charges:** Babich, "Dirty Business."

June 2004—Lost Mountain

p. 147 **The elk:** Scott Weidensaul, *Mountains of the Heart* (Golden, CO: Fulcrum, 2000), pp. 115–17.

p. 147 **According to Dave Maehr:** David S. Maehr, "What Follows the Elk?" *Wild Earth,* Spring 2001, p. 53.

p. 148 **David Ledford of the Rocky Mountain Elk Foundation:** Art Jester, "Wildlife, Coal Advocates Hold Talks," *Lexington Herald-Leader,* June 28, 2005, p. B1.

p. 150 **"Why you goddamn . . . *beatnik*":** Gurney Norman, *Kinfolks: The Wilgus Stories* (Frankfort, KY: Gnomon Press, 1977), pp. 69–70.

p. 152 **A few years ago:** Roger Alford, Associated Press, August 6, 2004.

p. 152 **Silas Miller has black lung:** Ibid.

p. 153 **A 2004 report:** "Twenty-Second Annual Evaluation Summary Report for the Regulatory and Abandoned Mine Land Reclamation Programs Administered by the Commonwealth of Kentucky for Evaluation Year 2004," U.S. Department of Interior, Office of Surface Mining (Washington, DC: GPO, 2004), p. 30.

p. 153 **Of the 5,825,756:** "Answers to the 10 Most Frequently Asked Questions." htpp:// www.osmre.gov.

p. 155 **"The coal-output of the world":** Henry Adams, *The Education of Henry Adams* (Boston: Mariner, 2000), p. 490.

p. 155 **"At the rate of progress since 1800":** Ibid.

The Ecovillage

p. 157 **The Union Army showed them its gratitude:** Marion B. Lucas, *A History of Blacks in Kentucky* (Frankfort: Kentucky Historical Society, 1992).

p. 158 **In his most recent book:** David W. Orr, *The Last Refuge* (Washington, DC: Island, 2004), p. 29.

July 2004—Lost Mountain

p. 168 **fifty-three were killed:** Brandon Ortiz, "Coal-Hauling Roads: Twice as Deadly," *Lexington Herald-Leader,* January 16, 2005, pp. A1f.

p. 170 **As the wife of one trucker said:** Tom Hansell, *Coal Bucket Outlaw.* Appalshop Films, 2002.

p. 172 **Berry begins:** Wendell Berry, "Two Minds," *The Citizenship Papers* (Washington, DC: Shoemaker & Hoard, 2003), pp. 85–106.

pp. 172–173 **To think with the sympathetic mind:** Aldo Leopold, *A Sand County Almanac* (New York: Ballantine, 1970), p. 137.

p. 173 **"Only the mountain":** Ibid.

Which Side Are You On? (Part 3)

p. 174 **The Harlan County Coal Operators Association paid:** John W. Hevener, *Which Side Are You On?* (Champaign-Urbana: University of Illinois Press, 2002), p. 40.

p. 174 **The decade-long struggle:** Ibid., p. 33.

p. 175 **On May 5:** Ibid., p. 35.

p. 175 **In his fascinating memoir:** G. C. Jones, *Growing Up Hard in Harlan County* (Lexington: University Press of Kentucky), pp. 48–49.

p. 176 **In a three-month period:** Romalis, *Pistol Packin' Mama,* p. 80.

p. 176 **She earned the moniker:** Ibid., p. 81.

p. 177 **Finally, on February 9, 1937:** Hevener, *Which Side Are You On?,* pp. 135–36.

p. 177 **Kentucky governor:** Ibid., p. 142.

pp. 177–178 **In addition, President Roosevelt:** Ibid., p. 153.

p. 178 **G. C. Jones remembered:** Jones, *Growing Up Hard in Harlan County,* p. 138.

p. 178 **But a declining demand for coal:** Hevener, *Which Side Are You On?,* p. 183.

p. 179 **When the first wave of protesters:** Jim Jordan and Beth Musgrave, "Union Miners Protest Bankruptcy Proceedings," *Lexington Herald-Leader,* pp. A1f.

RFK in EKY

p. 194 **In his recent book:** Peter Edelman, *Searching for America's Heart: RFK and the Renewal of Hope* (New York: Houghton Mifflin, 2001), p. 30.

September 2004—Lost Mountain

p. 210 **"One of the penalties":** Leopold, *A Sand County Almanac,* p. 197.

p. 210 **That the land *is* one organism:** Ibid., p. 190.

p. 211 **"Wisconsin not only had a round river":** Ibid., p. 188.

p. 211 **"one humming community":** Ibid., p. 193.

p. 211 **Of any biome:** Ibid., p. 198.

p. 213 **"reconstitute the natural world":** Václav Havel, "Politics and Conscience," *Living in Truth* (London: Faber and Faber, 1989), p. 149.

NOTES

Conclusion

p. 226 **Furthermore, Diamond wrote:** "The Last Americans," *Harper's,* June 2003, p. 51.

p. 227 **Twenty years ago:** Wangari Maathai, *The Green Belt Movement* (New York: Lantern, 2003).

p. 229 **fifteen hundred acres have been reclaimed:** Paul F. Rothman, as cited in "Testimony to the Commonwealth Energy Policy Task Force Appalachian Regional Reforestation Initiative," document issued by the Kentucky Department of Natural Resources, November 15, 2004.

p. 232 **"we are left with no expansion of wisdom":** Alfred North Whitehead, *Science and the Modern World* (New York: Mentor, 1925), p. 176.

p. 233 **"His consistent theme":** Ibid., p. 79.

p. 234 **"Love the quick profit":** Wendell Berry, *Selected Poems* (Washington, DC: Counterpoint, 1998), pp. 87–88.

p. 234 **38 percent larger:** Marilyn Berlin Snell, "Better Homes and Garbage," *Sierra,* January–February 2005, p. 28.

p. 238 **"To say that God":** Stuart Hampshire, *Spinoza* (Baltimore: Penguin, 1951), p. 44.

p. 239 **Two billion years ago:** Lynn Margulis and Dorion Sagan, *Microcosmos* (Berkeley: University of California Press, 1997), p. 117.

p. 239 **"Not only is Dr. Margulis's theory":** Richard Dawkins, *River Out of Eden* (New York: Basic, 1995), p. 46.

pp. 239–240 **"We cannot fathom":** Charles Darwin, quoted in Margulis and Sagan, *Microcosmos,* p. 32.

p. 241 **"In short":** E. O. Wilson, *In Search of Nature* (Washington, DC: Island, 1996), p. 166.

p. 241 **"The basis of any real morality":** Quoted in Donald Worster, *Nature's Economy* (Cambridge, England: Cambridge University Press, 1977), p. 186.

p. 242 **Frank Lloyd Wright:** Frank Lloyd Wright, *The Living City* (New York: Horizon, 1958), p. 46.

Recommended Reading

Berry, Wendell. *The Citizenship Papers*. Washington, DC: Shoemaker & Hoard, 2003.

———. *A Timbered Choir: The Sabbath Poems, 1979–1997*. Washington, DC: Counterpoint, 1998.

———. *The Unsettling of America*. San Francisco: Sierra Club Books, 1977.

Billings, Dwight, Gurney Norman, and Katherine Ledford, eds. *Back Talk from Appalachia*. Lexington: University Press of Kentucky, 1999.

Caudill, Harry. *My Land Is Dying*. New York: E. P. Dutton, 1971.

———. *Night Comes to the Cumberlands*. Boston: Little, Brown, 1963.

RECOMMENDED READING

Constantz, George. *Hollows, Peepers and Highlanders: An Appalachian Mountain Ecology*. Missoula, MT: Mountain Press Publishing, 1994.

Drake, Richard B. *A History of Appalachia*. Lexington: University Press of Kentucky, 2001.

Giardina, Denise. *Storming Heaven*. New York: Ballantine, 1988.

Maathai, Wangari. *The Green Belt Movement*. New York: Lantern Books, 2003.

Johanssen, Kristin, Bobbie Ann Mason, and Mary Ann Taylor-Hall, eds. *Missing Mountains*. Nicholasville, KY: Wind Publications, 2005.

Norman, Gurney. *Kinfolks: The Wilgus Stories*. Frankfort, KY: Gnomon Press, 1977.

Still, James. *From the Mountain, From the Valley: New and Collected Poems*. Lexington: University Press of Kentucky, 2001.

———. *River of Earth*. Lexington: University Press of Kentucky, 1978.

© Mary Bolin-Reece

Erik Reece teaches writing at the University of Kentucky in Lexington. His work appears in *Harper's*, *Orion*, and *The Oxford American*, among other publications. He was the recipient of the Sierra Club's David R. Brower Award, and his *Harper's* story on which *Lost Mountain* is based won the Columbia University School of Journalism's 2005 John B. Oakes Award for Distinguished Environmental Journalism.